HITCHHIKERS' GUIDE TO ELECTRONICS IN THE '90s

by David Manners

For Sash and Gem

Published by Computer Weekly Publications,
Quadrant House, The Quadrant,
Sutton, Surrey, SM2 5AS

Publications Manager: John Riley
Deputy Publications Manager: Robin Frampton
Publications Assistant: Katherine Canham

 REED BUSINESS
PUBLISHING   © 1990
GROUP

A British Library Cataloguing in Publication Data
Manners, David
    Hitchhikers' Guide to Electronics in the '90s
    1. Electronics industries
    I. Title
    338.47621381

ISBN 1-85384-020-3

Printed by Hobbs The Printers, Southampton

CONTENTS

networking and the decline of the mainframe; the rise of the cordless telephone; the pre-eminence of software; transportability; technological democratisation.

How to become a millionaire; the developments in the key component technologies for the Nineties - chips, screens, batteries and CD-ROMs. What the technology/price trends are; how to predict them; how to exploit the 'learning curve'.

Will the shrinking transistor continue as the main force? Will superconductivity take over? Will bio-technology take over? Will Japan take over the world computer industry? What would be the implications for entrepreneurism? Big business versus individuality. Can the US regain technological superiority? Can Europe re-build its technological and industrial strength? Will Eastern Europe become a significant market/producer?

FOREWORD

The strides made by electronics technology in the last 40 years would have been scarcely conceivable by the electronics engineers of the Forties.

Imagine, a transistor, never mind a million bits per transistor, a computer which sits on your lap and does not take up half the space of the Albert Hall, a machine which can transfer documents around the world in seconds; these are giant strides.

In this book David Manners puts these advances in an historical context, looks at the microchip technology which is at the heart of all technological advances, and surveys the major industrial electronics power houses. He looks into the future of both the technology and emerging markets.

If you want to know the origins of electronics, its present and its future, this is the book for you.

Mick Elliott
Editor of *Electronics Weekly*

CHAPTER 1

WHAT'S INTERESTING ABOUT ELECTRONICS?

The potential to reproduce human functions in silicon; the democratising effects of technology; electronics goodies of the Nineties; 'who wants to be a millionaire?'; the international race for electronics leadership.

Electronics offers a way to make an artificial you. That's just one thing that's interesting about it.

Already electronics could make a nose which is as good as your nose - but it would be the size of a fridge; or an ear which is as good as your ear - but it would be the size of a wardrobe; or an eye like your eye - but it would be the size of a car; or a brain as retentive as your brain - but it would be the size of a house.

Electronics may be able to offer you a kind of immortality - so that the information inside your skull could exist forever

What electronics will be able to do sometime in the twenty-first century is to make a nose the same size as your nose; an ear the same size as your ear; an eye the same size as a human eye; a brain that will fit in a human skull and perform the functions of taste and touch.

In other words it will reproduce the five senses and the brain. Moreover the material used - silicon - lasts forever. That means someone could say to you: 'Transfer all your knowledge and experience into our artificial brain'.

If you accept the offer then you - if what is inside your skull is you - could exist forever. So electronics may be able to offer of a kind of immortality.

Fanciful? For years American defence scientists have been working on ways of plugging add-on information packs direct into the brain. By plugging such packs into the brains of monkeys, monkeys have been able to perform complicated tasks like piloting missiles.

Sometime in the next century they may be able to reverse that process - i.e. transfer the contents of a human brain into an artificial brain. Such a brain could be reproduced in its thousands and sold.

People would then be able to plug into the contents of someone else's brain. Ever wondered what it would be like to be plugged into the brain of Mrs Thatcher or Michael Jackson?

Scary? For the people of the eighteenth century it would have been scary to think that you could see people walk about and hear them talk when they are absent or dead. Film and TV has made this possible.

Nowadays it seems equally odd that we may be able to tune into the thoughts of someone who is absent or dead. But to the citizen of the twenty-first century it will be as commonplace as films and video recordings are to us.

Consider the advantages we would have gained if all the accumulated brain-power and experience of the great men and women of history was still available to us

However the artificial brain won't be just a spooky trick, it will have advantages. Consider the advantages we would have gained if all the accumulated brain-power and experience of the great men and women of history was still available to us.

Think what it would be like to tap into the brains of Einstein and Shakespeare. All those people of knowledge and wisdom would remain at the service of the world. The motivations of a Napoleon, the passions of a Nelson but also the evils of Hitler would be known to us all - undiluted by the embellishments and sanitations of autobiography and history.

Products of the Nineties

Even if you don't want to buy immortality in the electronics supermarkets of the next century, just in the next decade there will be some interesting products on offer.

For instance, sometime during the Nineties, if you want the knowledge of a doctor you'll be able to buy it in a pack and plug it into your TV or personal computer. If you want a whole library - you'll buy a library pack for a few pounds - and have the contents of thousands of books available at the touch of a button.

THE DEMOCRATISING EFFECTS OF TECHNOLOGY

Another interesting contribution coming from electronics, and one which may be more difficult to believe in than immortality, is that it could make governments honest.

When everyone has a portable cordless phone costing the same as a calculator, videocameras get so small they can be put in a pocket, and fax machines are cheap and cordless - all of which is coming in the next decade - then the scale of global word of mouth communication will make it difficult for governments to lie to their people through the 'official' media.

It is often the mass belief in lies that sustains tyrannies - in offering cheaper and easier communication of pictures, data and speech, electronics may be a force for freedom in the world

Since it is usually the mass belief in lies that sustains tyrannies, it is possible that the effect of electronics in offering cheaper and easier communication of pictures, data and speech will be a force for freedom in the world. For instance TV broadcasting from satellites persuaded many East Europeans that they could live a better life in Western Europe - with the electrifying effects of November and December 1989. More direct person-to-person contacts via increasingly portable electronic equipment will accelerate the global dissemination of word of mouth information.

Electronics may also be a force which prevents the worst excesses of governments. TV pictures of the Vietnam War are said to have brought US public opinion out in sufficient force to stop the war. Televising of Tiananmen Square had severe political consequences for the government which authorised the massacre. Instant communications between people around the world will increasingly prevent the cover-up of atrocities by governments.

ELECTRONICS PRODUCTS OF THE FUTURE

The electronics industry's message is: 'You ain't seen nothing yet!' If you thought pocket TVs, pocketable calculators and telephones, multi-rhythm keyboards and portable computers were clever tricks - you are going to be amazed by all the new goodies electronics is going to give you.

Every year the price will drop and every year electronics products will be able to do more and more

Give might be the wrong word. You'll have to pay - but every year the price will drop and every year the products will be able to do more and more.

So what are some of the things people have already thought of that will be the products of the electronics industry in the next decade?

The Products of the Next Decade?

- cars that drive themselves;

- non-crashable cars;

- domestic robots;

- a machine the size of a paperback that will hold the contents of all the books in a library and display them for you to read;

- a wristwatch-sized portable communicator that will allow you to talk to anyone in the world from anywhere in the world;

- machines which will automatically translate what you say simultaneously as you say it;

- computers and portable telephones as cheap and plentiful as calculators;

- TVs so light and small that they'll fit in your shirt pocket;

- pocketable fax machines;

- remotely operated automated factories;

- voice operated domestic appliances - fridge/alarm clock/ bath taps/lights; lawnmowers and hoovers which can 'see' using artificial vision to allow them to turn corners and avoid obstacles.

All the products in this list are basically better ways of doing things than we already have. This is one function of the electronics industry, but there is another more interesting and much more profitable function - making something that did not exist before.

PROFITS IN ELECTRONICS

For instance making a TV today is not a very profitable exercise, but it was immensely profitable to make the first TVs. In the next twenty years there could be a lot of new things like that and the real money to be made over the next twenty years will be in thinking up entirely new things which the technology can do.

That's another interesting thing about electronics - it could make you very rich during the next decade. More and more the electronics companies look around for things to do with the technology they've got.

> The real money to be made over the next twenty years will be in thinking up entirely new things which the technology can do

If you can understand the main trends in component technology, realise what is going to become possible a few years out and think up a product that fits the increasing capabilities of the technology then you are going to make a lot of money between now and AD2000 whether you make your idea yourself or sell it to an electronics company.

Millionaires Overnight

If the past is anything to go by, many people will come up with such ideas because another of the interesting thing about the electronics industry - is that it's continually creating 'overnight' millionaires.

For instance, a couple of American students Steve Wozniak and Steve Jobs founded and built the personal computer company Apple into a $5bn company within a decade. Sir Clive Sinclair became a multi-millionaire on one product in a few years. Alan Sugar built Amstrad into a half-billion pound company in less than ten years.

In Japan, Konosuke Matsushita, who died in 1989, built the $40bn Matsushita Electrical Industries up from nothing within his own lifetime and Akio Morita founded the Sony Corporation as a young man in 1945 and has seen it grow into a $7bn company with a string of innovations from consumer tape recorders to the Walkman.

But you don't have to build a huge company to make your fortune, creativity is the key to electronics millionairedom. Teenagers make fortunes thinking up computer games. Designers of microchips in their early twenties can write their own pay cheques. Mushrooming software companies propel their founders to millionairedom in a couple of years.

That is going to stay the norm in this still very immature industry. While the technology involved in the design and manufacture of electronics products is still evolving - and there is still no sign of an end to its evolution - there is always opportunity for new people, new ideas, 'right-angled' thinking, leaps of imagination and wild eccentricity.

Electronics technology has not been sewn up and taken over by large companies; it is still the province of the inventor, the entrepreneur and the person with a new idea

Electronics technology is not something that has been sewn up and taken over by large companies; it is still the province of the inventor, the entrepreneur and the person with a new idea.

THE INTERNATIONAL RACE FOR ELECTRONICS LEADERSHIP

The history of the second half of the twentieth century is very much tied into the electronics industry's development. World politics is a game of wealth and power in which technology leadership is a key element.

World politics is a game of wealth and power in which technology leadership is a key element

The world's leaders became acutely aware of the importance of technology during the 1939-45 war. Radar, encryption (coding/decoding) machines, rockets, aircraft and atom bombs woke politicians up to the importance of having a technical lead. By the end of the war, knowing that mastery of space technology would provide economic and military power for the rest of the century, Russia and America raced to Berlin to grab as many German rocket scientists as they could. Russia got there first and grabbed the best.

At the same time the Japanese were pondering that the reason they had lost the war was because of American technological superiority demonstrated by the atom bomb. They set out to achieve superiority in electronics technology, knowing that such a position would give them the edge in the new war - the war to win in world commercial markets.

The Race for Space

So the Russians went for rocket technology, the Japanese went for electronics technology, and the Americans went for a bit of both. Then came the 'Sputnik' the Russian spacecraft which was the world's first manned space flight.

The shock to the American psyche of being beaten in a technological race was so great that US President John F. Kennedy was able to ask the US Congress for billions of dollars to put a man on the moon by the end of the Sixties - and got it. But the Americans rockets were not as powerful as the Russian rockets, so the Americans tried a different tack. They aimed to make the rockets' guts - the instruments, the control equipment, the communications gear - smaller and lighter. By doing that the weaker American rockets would still do the same job as the more powerful Russian rockets.

The Americans' motto 'In God We Trust' didn't let them down. The early Sixties was a time when a revolution in the ability to miniaturise electronics equipment was just getting started. It was the beginning of the silicon chip, sometimes called the microchip and known to the industry as the 'integrated' circuit or just 'IC'.

The first microchip had been made by Jack Kilby working for Texas Instruments in October 1958. Basically what it did was get rid of wires. Instead of using a wire to connect one part of an electronic product to another part, it plumped both parts down together on the same piece of material with a fixed connection between them. Its

The first microchip, by getting rid of wires, allowed infinite possibilities for miniaturisation

advantage was that it allowed infinite possibilities for miniaturisation.

This is exactly what the moonshot guys wanted. If they could make the electronics equipment in their rockets smaller and less power consuming then their small rockets could put into orbit the equipment necessary to perform the same tasks as the larger Russian rockets.

The Expansion of the Electronics Industry

So, a large chunk of the billions of moonshot dollars went to the electronics industry. Much of it went to the big US electronics companies like the computer manufacturing company IBM (known as 'Big Blue') and the telephone equipment manufacturing company AT&T (known as 'Ma Bell').

Big Blue, Ma Bell and many others passed on research and development (R&D) contracts to any company big or small which was prepared to pioneer the new microchip technology. This was to have a major impact in the way the chip industry developed.

The flood of dollars soon produced results. The effect of mobilising the US electronics industry from the largest corporation down to the one person entrepreneur was electric. Between 1960 and 1970 over 40 new companies started in California to pioneer microchip technology development.

It was a good time to be an electronics engineer. With a development contract from a large company you could kiss your employer goodbye, set up a company on borrowed money, develop the technology required by the contract and, as a sideline, use the technology you had developed to make some commercial products of your own.

The microchip companies flocked together in California - eventually there were so many of them that their locality was named 'Silicon Valley'

Millionaires blossomed like a desert after drought. Usually they blossomed in California where the first chip company Fairchild Semiconductor set up shop. The microchip companies flocked together to feed off each others' ideas and poach each others' employees. Eventually there were so many of them that their locality became called 'Silicon Valley' after the basic material they used to make their chips.

It all worked out fine. The Americans got their man on the moon by the end of the Sixties; a lot of technologists got very rich, Silicon Valley kept on expanding with over 100 more new microchip companies founded in the Seventies and 150 more in the Eighties and, while the Russians still kept their lead in rocket technology, the Americans took a decisive lead in microchip technology.

The Japanese Threat

But across the Pacific from the San Francisco peninsula, where the American chip companies clustered in their Valley, a threat was stirring. The Japanese were rebuilding their industrial base. Raw materials were non-existent in Japan and agriculture was no export earner for so tiny and mountainous a country. Manufacturing was Japan's only way of earning a place in the world economy. Japan's manufacturing extended on many fronts - car and motorbike manufacturing, ship-building, steel manufacturing - but electronics was the coming industry and in that industry Japan made a specially concerted effort.

Electronics was the coming industry and in that industry Japan made a specially concerted effort

In making their play for electronics markets, the Japanese watched America like a hawk and the Americans, determined during their military occupation of Japan to reshape the country as a modern, classless, industrial democracy, encouraged Japanese business leaders to visit America, see American factories and get insights into American technology.

The Success of Japan

It worked like a charm. With growing confidence the Japanese took advantage of this access. Profits from the mature industries like ship-building and steel were ploughed back via the redoubtable Ministry for International Trade and Industry (MITI) into the rising industries of consumer electronics, computers and telecommunications.

But the Japanese did not copy everything. Although they also recognised that the key enabling technology for the electronics industry was microchip technology with its ever-increasing capability to miniaturise, they didn't go down the entrepreneurial route of the Americans. In Japan it was the big companies, directed and funded by MITI, which developed chip technology.

In 1971 MITI made available $600m over four years to the six largest electronics companies just to develop microchip technology. In 1976 it followed that up with another $325m over five years. It did the trick. In late 1983, the Japanese brought out a new type of advanced chip and it was a year before the Americans could match it. Five years on from there, Japan was outproducing America in chips; Japan had won.

Reasons for Japan's Success

There were two reasons why Japan won. The first was because it was ruthless in its marketing. Japanese marketing philosophy puts market share as the number one goal and other considerations like profitability a long way second. In practice that meant that they sold the chips in foreign markets for pretty well anything they could get for them.

The second reason Japan won in the microchip market was because although the small US companies could develop the technology with government money and run with it to market very effectively, small companies could not generate sufficient profits from the notoriously unstable microchip market to fund the R&D for the next generation of microchip.

Japan, with huge companies involved in every field of electronics from computers to telecommunications, had the revenues to keep up R&D through good times and bad

The Japanese, however, with huge companies involved in every field of electronics from computers to telecommunications, had the revenues to keep up the R&D through good times and bad.

The significance of winning in the microchip technology and microchip markets, was that Japan got control of the means of production and supply of the electronics industry's building blocks. With that control it could start to put the squeeze on the equipment manufacturers of the West.

And it did. The US/Japan 'Chip Wars' of the late Eighties were the result of putting on that squeeze. Ironically, they started as a result of the Americans accusing the Japanese of dumping chips. The Japanese riposte was to restrict supplies of chips so causing prices to soar. American equipment manufacturers then screamed blue murder as they were unable to make equipment or were forced to raise prices to uncompetitive levels.

The result was that Japanese chip producers made enormous profits which went into R&D to extend their technological leadership.

THE IMPORTANCE OF ELECTRONICS

The Current State of Play

So, at the start of the Nineties, we now have a world where the Russians lead in rocket and space technology, the Japanese lead in microchip technology, and the Americans lead in military and computer technology and share with the Europeans leadership in telecommunications technology. All of which is important for three reasons:

- the electronics industry is now the largest industry in Japan and will probably become the world's largest industry in the Nineties or early 2000s - and whoever dominates electronics markets will be powerful and rich;

- military spending is set to drop around the world and as it does so the West's electronics industry will have to generate more of its revenues from the market and less from selling to governments;

- markets will be won by the companies with the best enabling technology and the most important enabling technology in electronics is microchip technology.

Which is why even the home of laissez-faire markets, America, is injecting the tax-payer's dollar into the chip R&D activities of commercial companies. It is why the EEC is beginning to inject serious money (several billion dollars during the Nineties) into commercial chip R&D and it is why the rising electronics nations of Asia - Korea, Taiwan and China - are stepping up their investments in developing the technology.

So there are a few things interesting about electronics: it offers routes to immortality or mere millionairedom; it explains the history of this century and suggests the pattern of the next; it decides the economic fate of large chunks of the globe like Europe, America, Japan and Asia; and it offers limitless opportunities for leisure and pleasure.

CHAPTER 2

WHAT'S IT ALL ABOUT?

From Cavendish to Faraday; from Edison to Kilby; from the vacuum tube to the transistor to the chip; the Americans invent the chip industry; Moore's Law.

Many schoolchildren know that:

- an electron revolving round a core, like planets around the sun, is what makes up an atom;

- the ancient Egyptians discovered static electricity by rubbing amber with silk;

- static electricity is caused by electrons becoming detached from their orbits around the atomic core.

Many schoolchildren don't know that:

- in 1600 an Englishman, William Gilbert, coined the term 'electricity' for the phenomenon of static electricity and observed that it occurred with many common materials besides amber and silk. He also noted that it was somehow connected with magnetism.

- in 1729 another Englishman, Stephen Gray, pointed out that static electricity could pass through certain materials and could therefore be transferred from one place to another along strips of material. This phenomenon he called 'conducting'. A material through which electrons could travel was called a 'conductor'.

- in the mid-1770s the English eccentric genius Henry Cavendish, who had two Dukes for grandfathers, found a way of measuring electricity flowing through conductors. Cavendish, a recluse who hated women so much that he insisted his female servants be kept out of his sight, would hold a wire in each hand so the electric current passed through his body. This allowed him to measure the strength of the current by feeling how far up his arms the pain went.

- in Italy in 1790 Luigi Galvani related how static electricity manufactured on one side of his laboratory had caused a dissected frog's leg to twitch on the other side. Realising that the electricity had been 'conducted' across his lab, Galvani then connected up the frog's muscle and nerve with a metal connection and noticed the same twitching. Obviously electrons were passing between the nerve and the muscle. The phenomenon was named after him as a 'galvanic' flow of electricity.

Schoolchildren progressing to the elementary physics courses will soon learn about:

- Alessandro Volta of Italy who made the first battery in 1800 and had 'volts' named after him;

- Andre-Marie Ampere of France who demonstrated some magnetic effects of electric currents and gave his name to the electrical measure 'amps';

- George Ohm of Germany who formulated the scientific law behind the phenomenon and had 'Ohm's Law' named after him.

The Brilliance of Faraday

But most people will have learnt how the greatest of the electrical pioneers, Michael Faraday of England, showed that all these kinds of electricity - galvanic, static, magnetic and voltaic - were simply different manifestations of the same force.

Faraday showed that all kinds of electricity - galvanic, static, magnetic and voltaic - were simply different manifestations of the same force

Faraday then demonstrated the key turning point in the science - that electrical force produces magnetic force and vice versa. That meant you could convert magnetic force to electrical force and electricity back into magnetism.

The importance of the discovery was that it meant you could use mechanical energy to generate electrical energy. Faraday showed that by rotating a copper disc between the attracting and repelling ends of a magnet you could make electric current. That was the key which opened the door leading from the mechanical age into the electrical age.

The discovery - that by rotating a copper disc between the attracting and repelling ends of a magnet made an electric current - opened the door from the mechanical to the electrical age

That made possible the generator or dynamo which used mechanical power to produce electricity. In 1821 Faraday went further and showed the effect in reverse i.e. that you could make a machine into which you could pump electricity at one end and get mechanical energy - motion - out of the other end. In other words, he showed that you could make an electric motor.

The Genius of Edison

Faraday did more than open the door to the electrical age - he inspired its greatest inventive genius - Thomas Edison of America. The 21 year-old Edison found a copy of Faraday's journals in an American bookshop. They were the spark that fired the American's imagination and ignited an unparalleled talent for invention and innovation.

Simultaneously with Sir Joseph Swan of England, Edison demonstrated that when you passed electricity through a filament of poorly conducting carbon you produced light - the light bulb. Edison was the original role model of the anarchistic technologist defying conventional scientific wisdom to achieve his breakthroughs. He set up what he called an 'invention factory' in the US. His employees loved and loathed him as he veered from the charmingly entertaining to the tyrannically bossy. As a self-taught man he laughed at the formally trained scientists and they in turn dismissed Edison as anti-intellectual.

Edison told his friends that the target of his invention factory was to come up with a minor invention every ten days and a 'big trick' every six months. In 1877 he performed a notable double - he invented both the gramophone and the microphone. Edison did not scoop the biggest prize however. Alexander Graham Bell of England patented the telephone in 1876.

Bell and the Telephone

For many years it had been possible to send signals down wires in the form of Morse Code and in 1850 a submarine cable across the Channel was laid for that purpose. However Bell's invention made it possible, for the first time, to send speech down wires.

Bell patented his invention in the US because it had an indifferent official reception in the UK. The English Post Master General at the time, John Tilley, commented: 'While the Americans might need such a daft thing, Britain still has plenty of small boys to run around with messages.'

THE BIRTH OF ELECTRICITY

With Edison and Bell electricity came into its own. Electrons pumped through bulb filaments resulted in light; pumped into a motor they resulted in motion; pumped into wire coils they resulted in heat; sent along wires they allowed communication between distant people.

With Edison and Bell electricity came into its own, bringing inventions for light, warmth, energy and communication

Light, warmth, energy and communication are basic human needs and very saleable. The inventions were the seeds of a vast new industry.

The use of electricity to send signals through the air (radio) was also a nineteenth century invention. It had been observed that signals sent down wires caused 'interference' with other electrical equipment. Clearly some of the signals were escaping from the wires into the surrounding atmosphere. Ways were found to make the wire so as to maximise these escaping signals i.e. transmitters were invented.

Sir Ernest Rutherford of England transmitted radio signals for three quarters of a mile at Cambridge in 1895. In 1901, Guglielmo Marconi demonstrated trans-Atlantic radio transmission. So, by the end of the nineteenth century, all the basic products of the electrical age - light, telephone, radio and the electric motor - were invented.

Throughout the twentieth century people bought huge numbers of electrical appliances and the electric power which made them work

Throughout the twentieth century people bought huge numbers of electrical appliances and the electric power which made them work. Power stations were built to supply the ever-increasing demand for electric power and brains were racked to think up new ways in which that power could be put to use.

For most of the twentieth century the brains were fertile. They thought up: washing machines, irons, gramophones, fires, sewing machines, toasters, kettles, milk floats, cookers, air conditioners, fans, fridges, dishwashers, lamps, hairdryers, polishers, hoovers, drills, sanders, hedge trimmers, mowers, trains and more.

The list of the products of the electrical industry is long. Most of these products were the old mechanical products i.e. machines made up of cogs, levers and wheels which until then had been powered by steam, horse or muscle.

One thing the electrical industry did was to take these mechanical products and convert them to run on electricity. For instance a train could be converted from using a coal-fired water boiler pushing a piston to and fro to an electricity-powered electric motor turning a spindle.

However another class of products sprang up which were more than simply mechanical products converted to run on electricity. These were products like telephones, microphones and lightbulbs. These can be considered the true products of the electrical age because they could never have been made by mechanical means.

THE AGE OF ELECTRONICS

Then came the age of electronics. The distinction between electrical technology and electronic technology (a distinction which has been argued about for many years) is the distinction between manipulating 'captive' electrons flowing along wires and manipulating 'free' electrons in vacuums, gases and other materials.

The tool which made it possible to manipulate 'free' electrons for the first time was the 'thermionic valve' or 'electron tube'. The prime mover behind this device was Edison.

As such Edison is the key figure in the transition from the electrical age to the electronics age just as Faraday was the major force behind the move from the mechanical age to the electrical age.

From the Vacuum Tube to the Transistor

In his work on light bulbs Edison had noticed that, in certain conditions, an extra current to the one used to produce light was created in the vacuum of the bulb. This became known as the 'Edison Effect' and its theoretical basis was first explained by Sir J. J. Thomson of England. In 1904 another Englishman, Sir John Fleming, constructed a vacuum tube that could reproduce the Edison Effect at will. The significance of Fleming's tube is that it created the conditions under which free electrons could be manipulated which meant that, for the first time, it was possible to manipulate electrons in other ways than simply sending them down wires.

Fleming's tube, for the first time, made it possible to manipulate electrons in other ways than simply sending them down wires

The tube made it possible to boost ('amplify'), diminish, generate, vibrate ('oscillate'), modulate and switch electrons. This is the area of activity known as electronics which is why Edison's perception and Fleming's invention are regarded as the keys to the electronic age. However the tube, which Fleming called a 'diode' because it had two 'electrodes' in between which the current flows, did not immediately perform the application for which Fleming had intended it to which was to act as a receiver for radio signals.

Then in 1907 Lee de Forest of the US improved on Fleming's device by adding a third electrode (making a 'triode') which 'amplified' or boosted the signals so that they were strong enough to permit the reception of speech via radio signals.

Until then radio signals had only been used to transmit messages in Morse Code as they were too weak to carry speech. The de Forest triode made it possible to make vacuum tubes powerful enough to allow the mass broadcasting of radio programmes (broadcasts started in the Twenties) and paved the way for television.

The vacuum tube became the key building block for making computers and, in time, for making every kind of electronic equipment

Moreover the capability of the tube to turn signals on and off very quickly (to 'switch') meant that electronic computers could be built. So Fleming's vacuum tube became the key building block for making computers and, in time, for making every kind of electronic equipment.

As with the transition from the products of the mechanical age to the products of the electrical age, it is possible to classify some electronics products as simply improved electrical products and other products as only becoming possible because of the newfound ways to manipulate electrons.

Examples of this latter kind of product are the broadcasting of speech and pictures, electronic computing and global communications via satellite. These are the true products of the electronic age. Broadcasting, telecommunications and computing have transformed the second half of the twentieth century. Huge industries have grown around them. That is the legacy of the tube.

As such, the more the tube could be improved on, miniaturised and made less power consuming, the more rapid would be the advances in the capabilities of the electronic equipment which it was used to build. The development of the technology for making tubes therefore became the key enabling technology driving forward the development of all electronics products.

However the tube was just a glass bulb enclosing a vacuum and it had many disadvantages: it was fragile, shortlived, bulky, power-consuming and generated colossal amounts of heat. If the future development of electronics products was to depend on the future development of tubes, the future development of electronics products looked like being a pretty slow business. As we now know, it wasn't.

From the Transistor to the Microchip

Why not? Very simply. Because the tube became replaced with a device made out of solid material which was capable of apparently infinite miniaturisation.

On December 23rd, 1947 three researchers working for Bell Laboratories, the research arm of the giant telecommunications company AT&T, invented a device made out of the material germanium which could, like a vacuum tube, amplify, oscillate and switch electrical signals.

The tube was replaced with a device made out of solid material which was capable of apparently infinite miniaturisation, the transistor

The three researchers were Walter Brattain, John Bardeen and William Shockley and their tube-replacement invention, called a 'transistor', brought them the Nobel prize.

AT&T's production arm, Western Electric, started producing commercial transistors for sale in 1951 and the following year it held an open symposium at which anyone could come and learn how to make transistors for the paltry entrance fee of $25,000.

Why did AT&T decide to license the technology of how to make transistors as widely as possible? There were three reasons: first to maximise the effort put behind the development of the technology; second to be seen to be performing a public service and so head off the anti-monopoly lobbyists (AT&T was continually criticised in the US for having a monopoly on the telephone system); third, to maximise revenues from licence fees.

There was one exception to Bell's requirement for licence fees. Transistors produced for use in hearing aids did not need a licence. This was a tribute to the work of Alexander Graham Bell in helping deaf people.

One of the earliest licensees of the transistor-making technology was a former lieutenant in the Japanese Imperial Navy called Akio Morita. After World War two Morita founded a consumer electronics company in Tokyo called Sony.

As often happens with breakthroughs, the industry establishment were slow to pick the transistor up or regard it as important

As often happens with breakthroughs in technology, the industry establishment were slow to pick the transistor up or regard it as important. During the early to mid-Fifties, transistors were more expensive than tubes and consequently there was no pressure to use them except in applications where miniaturisation was particularly important - as in hearing aids.

The companies which were first to take up manufacturing of transistors were the main tube manufacturers - General Electric of the US, Raytheon and RCA. As time went on, however, the manufacturing of transistors threw up new companies and the old leaders of the tube-making industry withered away.

A trend began - established industry leaders unable to respond to technical change were replaced by new companies

That was a trend which the industry was to repeat many times over the succeeding years - established industry leaders being unable to respond to technical change and new companies coming up to take their place. The process is still going on today.

The Silicon Transistor

A key development in the Fifties was the development of the silicon transistor. Up to 1954, transistors were made in germanium -a material which could not resist high temperatures. That ruled out its use in many products. Particularly it ruled the transistor out of use in military equipment.

At a conference of the Institute of Radio Engineers in 1954, various papers said that the answer to the problem of high temperature applications was to make transistors out of silicon. However, the technical difficulties of doing this, said the papers, would not be overcome for many years. Then up got Gordon Teal of Texas Instruments - a company involved in geological research which had only diversified into transistor manufacturing the previous year. Teal gave a paper describing how his company had made a silicon transistor.

The chief practical significance was that the US military started to take an interest. Once that was achieved, the development funds started to flow and the cost reduction process began in earnest. From 15 transistor manufacturing companies in 1953, the US industry had 26 by 1956. New companies and falling prices were to be the hallmark of the industry for the future.

The Dawn of the Microchip

Then came the next sensational breakthrough. The manufacturing of electronic products in the Fifties involved masses of wires. These were the only means of connecting up the different components like transistors, tubes, capacitors, inductors etc to produce the required 'circuit' to do the job which an electronic product is supposed to do.

During the Fifties people were beginning to see new possibilities. In 1952, Geoffrey Dummer of the Royal Radar Establishment (RRE) at Malvern gave a speech in Washington where he said that the next step after the transistor was to build 'electronic equipment in a solid block without connecting wires'.

In 1957, the RRE and Plessey demonstrated a model of Dummer's vision - a solid block of material with various components built into it - at the International Symposium on Components held in Great Malvern, Worcestershire. This was the first model of what became known as a 'solid state circuit' or 'integrated circuit' later to be popularised as the 'microchip' or just 'chip'.

In October 1958, a young scientist who had only joined Texas Instruments a few months before, tested an integrated circuit he had built containing transistors, resistors and diodes on a single piece of germanium. The circuit worked as planned. That successful test was the dawn of the age of the integrated circuit or 'microchip' as it became known.

Although made on the same piece of material the separate components of Jack Kilby's microchip had to be interconnected by wires. But, at the same time as Kilby was working on his microchip down in Texas, up in California Bob Noyce of Fairchild Semiconductor was working on a simple process by which different components on the same piece of material could be connected.

Noyce found a way of building the connections into the microchip using silicon as the material. Noyce's method of connecting up components was known as the 'planar process'. The effect of the planar process was to make the manufacturing of integrated circuits a commercially economical possibility.

After a considerable wrangle between Texas Instruments and Fairchild Semiconductor about who should hold the patent for the first microchip, it was agreed that a joint patent for the invention of the integrated circuit should be issued to Kilby and Noyce.

The invention of the chip turned the pace of electronic advance from a fast trot to a helter-skelter dash

The invention of the chip turned the pace of electronic advance from a fast trot to a helter-skelter dash. That was because ways were found to make the components on the chip continually smaller which meant that you could put more and more components on the same-sized piece of silicon.

Moore's Law and the Future

In 1964 Dr Gordon Moore of Fairchild Semiconductor observed that the number of components contained in the most up-to-date microchip doubled every year. The observation has held true ever since and the phenomenon has been dubbed 'Moore's Law'.

In 1964 Moore observed that the number of components contained in the most up-to-date microchip doubled every year - the observation has held true ever since

The reason why the operation of Moore's Law has had such an extraordinary effect on the capabilities of electronics products is simple: By the end of the Sixties the semiconductor industry could put 64 transistors on a chip. By the end of the 1980s decade the semiconductor industry was putting four million transistors on a chip. But the cost of manufacturing a chip has remained about the same whether it contains ten components or a million.

The cost of making a transistor on a chip fell from $2 in 1968 to one hundredth of a cent in 1985 - a decrease in cost by a factor of 200. The process is accelerating. That process means that you, the user of electronics goods, are going to get continually and rapidly decreasing prices. The price declines of the first products of this revolution - watches, calculators and personal computers have been apparent to everyone and the primary reason for those price declines is the operation of Moore's Law.

The sensational aspect of all this is that only now is Moore's Law starting to get serious. When you're doubling up from 1000 to 2000 transistors on a chip it doesn't make all that much difference to the end product. But nowadays when you're doubling up from four million to eight million you're talking about serious advances in the capability of the products into which the chips are going.

The industry reckons it will be putting one thousand million transistors on a chip by the year 2000 - that will allow you to replicate in silicon the workings of an eye, a tongue, a nose or an ear. The chip industry is just beginning to fulfil its potential - so watch out!

CHAPTER 3

WHAT DOES IT ALL MEAN?

What the jargon means: actives/passives; chips/ discretes; analogue/digital; bits/bytes; bipolar/ MOS; memory/logic.

The last two chapters hopefully have given a flavour of what the electronics industry is all about. They should have got over the message that the electronics industry makes the bits and pieces - the components - which are used to make up the products of other industries such as the computer industry, the telecommunications industry (making telephone exchanges, telephones and running telephone networks), the defence and avionics industry (making weapons, rockets and planes) and consumer products (hi-fis, calculators, TVs etc).

To many people the phrase 'electronics industry' encompasses all these industries. To others the term electronics industry means simply the components industry because that is 'pure' electronics whereas the other industries require other skills and disciplines.

However you use the term, it should be clear from the last two chapters that it is the advances in the components industry which make possible the advances in all other branches of the electronics industry.

It is the advances in the components industry which make possible the advances in all other branches of the electronics industry

To understand the industry, it's as well to look at some of the words which are used to describe the components and what they do. Like all professions electronics has its own odd language.

CIRCUIT

An electrical/electronic product consists of a 'circuit' or a series of circuits. One way to describe a circuit is as an obstacle course for electrons. As they flow along a wire you can boost the electron flow, reduce it, stop it, restart it, store it etc. This is done by inserting components in the wire.

For instance it is well known that the flow of electrons weakens the longer they travel along the wire. That is because the electrons meet 'resistance' in the material of the wire. Therefore if we want to weaken the flow wc use a very long wire and because it's too bulky to use stretched out wires we use coiled up wires. Such a coil is called a 'resistor'. Solid blocks of materials like carbon can achieve the same effect.

A circuit can be described as an obstacle course for electrons

Some people also know what a switch is. It can either make a break in the wire so cutting off the flow of electrons, or it can act like the points of a railway track and send them off down an alternative wire.

When the switch is operated by electrical current rather than mechanically it is called a 'relay'. When a magnetic field is wanted a component called an 'inductor' is placed in the wire. When it is necessary to store an exact amount of electricity at some stage along the wire it is done with a 'capacitor'.

To put all these in a 'circuit' to do something useful might result in something like this:

- you push the doorbell;

- the pressure mechanically 'switches' a circuit connecting it up making the electrons flow;

- because the mains current is too strong it is reduced by running it through a resistor;

- to ensure the current is at exactly the right strength in the right place along the line it is stored in a capacitor;

- the current then flows into an inductor which produces a magnetic field;

- the magnetic field pulls back a lever on a spring; the inductor is then turned off by a relay;

- the released lever strikes the bell.

ACTIVES AND PASSIVES

Components are split into two broad segments: actives and passives. Active components like the vacuum tube and the transistor contain the power to generate and alter electrical signals. Active components amplify (increase), oscillate (vary), rectify (convert ac to dc) and modulate (modify) electrical signals. Amplifiers, rectifiers, modulators and oscillators are all classed as active devices. Passive components consist of resistors, capacitors and inductors. All the components in the circuit for the doorbell were of the passive kind. The usual distinction between actives and passives is that the former introduces 'gain' (i.e can increase a given signal) whereas a passive component cannot.

The usual distinction between active components and passive components is that actives introduce 'gain' whereas passives cannot

CONDUCTORS, INSULATORS AND SEMICONDUCTORS

Conductors are materials along which electricity runs. Insulators are materials which stop electricity running.

Semiconductors are materials which can be treated so as to allow electricity to run through some parts of them but to stop electricity running through other parts.

Semiconductors are divided into two types:

Integrated Circuits and Discretes

As we've already seen an integrated circuit is a single piece of material with a number of components like transistors, resistors and capacitors made on it. An integrated circuit is also called an 'IC', a 'microchip', a 'monolithic chip' or just a 'chip'. Where transistors, resistors, capacitors and other components are made as individual separate items they are known as 'discretes'.

Analogue and Digital

'Analogue' which is used interchangeably with the word 'linear' means a 'real world' electrical signal like a measure of temperature, noise or weight. 'Digital' is used to describe an analogue electrical signal which has been converted into binary code. The reason for doing that conversion is that it is generally quicker to process and communicate digital signals.

An analogue/linear component is one which handles, accepts, processes or evaluates a signal from the real world

Binary code uses the two digits '0' and '1' to form a language for storing information conveyed by electrical signals.

So an analogue/linear component is one which handles, accepts, processes or evaluates a signal from the real world like the signal you get from an instrument that measures temperature, noise or weight. Analogue/linear components perform sensing functions i.e. in sensing temperature, vibration, humidity, noise levels etc. Analogue/Linear componentry contrasts with digital components. Digital components accept, process and store the information conveyed by electrical signals after that information has been converted into binary code.

This conversion is done by a device called, not surprisingly, 'a digital to analogue converter' (D/A converter). It is converted back to digital by an A/D converter.

An example is the difference between the analogue world of the gramophone record where the friction of a needle in a groove produces the music, and the digital world of the compact disc where the music has all been stored in the form of '0's' and '1's' inside the CD.

It is possible to have mixed linear/digital microchips where one part of the chip handles linear signals and the other half handles digital signals.

BITS AND BYTES

Bit is short for 'binary digit'. A binary digit is either a '0' or a '1'. Using combinations of '0' and '1' it is possible to code speech, pictures and data. For instance the figure 2 in binary code becomes 10; the figure 3 becomes 11; 20 becomes 10100. Once coded in binary form the speech/data/picture can be transmitted down telephone lines, broadcast or stored.

The '0' or '1's are represented in an electronic device by whether the device is conducting electrons or not. For instance if it is conducting it represents a '1', when not conducting it represents a '0'. This process of either conducting or not conducting is known as 'switching'. An electronic device that does it is called a 'switching' device. Obviously the faster such a device can switch the more binary digits or bits it can store.

Therefore a device's storage capacity is measured in bits e.g. a 1,000 bit memory can store 1,000 binary digits. A byte = 8 bits. The computer industry tends to talk in bytes and the electronics industry tends to talk in bits.

The computer industry tends to talk in bytes and the electronics industry tends to talk in bits

So discs, both floppy and hard, are usually measured in bytes or Kilo-bytes (one thousand bytes), Mega-bytes (a million bytes) or Giga-bytes (a thousand million bytes). But chips are usually measured in bits - K-bits and M-bits (G-bit chips have not yet been invented).

Although normally used to measure the memory storage capacity of a chip, bits are also used to refer to micro-processors. In this usage bits refers to the length of the chunk of information which a microprocessor can process at a time. So you get 4-bit, 8-bit, 16-bit, 32-bit and 64-bit microprocessors.

MILLI, MICRO, NANO AND PICO

These are not female Mexican quadruplets but degrees which can be applied to any parameter of measurement e.g. seconds, volts, amps, farads etc.

Milli = one thousandth
Micro = one millionth
Nano = one thousand millionth

Pico = one million millionth

So a picosecond is one million millionth of a second.

Hz, KHz, MHz, GHz

A Hertz is a measure of the speed of operation of many electronic devices. It indicates the number of cycles the device can perform in one second.

> A Hertz is a measure of the speed of operation of many electrical devices

2Hz = two cycles per second
Kilo-Hertz = one thousand cycles per second
Mega-Hertz = one million cycles per second
Giga-Hertz = one thousand million cycles per second

AMPS/VOLTS

Measures of electrical current.

FARAD

Measure of a capacitor's capacity.

BIPOLAR AND MOS (metal oxide semiconductor)

The two basic production methods for making microchips. Bipolar makes chips which work fast but use a lot of power.

MOS makes chips which can pack more components in than bipolar (i.e. 'denser') and use less power.

There are various forms of MOS - principally NMOS and CMOS. NMOS used to be the industry's workhorse production technology in the Seventies and early Eighties. CMOS or Complementary MOS is a particularly low-power MOS process. As of the end of the Eighties CMOS is the workhorse technology of the semiconductor industry.

BICMOS

A production process for chips which uses both bipolar and CMOS production techniques to make one chip. Theoretically it provides the speed of bipolar with the density and low power of CMOS.

LOGIC AND MEMORY AND LINEAR CHIPS

Logic chips perform the 'thinking' functions of an electronic product such as arithmetic or calculating functions. Memory chips store information. Linear chips perform 'sensing' functions measuring things like temperature, noise or weight.

Logic chips perform the 'thinking' functions of an electronic product

Standard ICs (integrated circuits) and ASICs (application specific ICs). Off-the shelf ICs are called standard ICs. ASICs are tailor-made.

RAMS AND ROMS

Random Access Memories (RAMs) are standard chips which store information but lose their memory when the power is switched off. Read Only Memories (ROMs) are standard chips which have the information in their memory fixed before they leave the factory. They do not lose their memory when the power is turned off.

RAMs are standard chips which store information but lose their memory when the power is switched off, whereas ROMs have the information in their memory fixed - even when the power is turned off

VOLATILE AND NON-VOLATILE

Volatile means that the chip loses its memory when the power is turned off. Non-volatile means it doesn't.

STATIC RAMS AND DYNAMIC RAMS

Both are memory chips which store information. Dynamic RAMs require a refresh electric current every few nanoseconds (in addition to their normal current) to keep their memory. Static RAMs don't need a refresh.

DRAMs can store one bit of memory using one transistor and one capacitor. SRAMs need four transistors and two capacitors to store one bit. SRAMs and DRAMs were both invented by Intel.

PROMS AND EPROMS

These are also memory chips. The programmable ROM is a ROM you buy from the factory and put whatever information you want into its memory.

Erasable PROM - a PROM you can erase by placing under ultra-violet light. The EPROM was invented by Intel.

E2PROMS AND FLASH

These are more memory chips. An electrically erasable PROM can be erased by an electric current instead of by ultra-violet light.

Flash EPROM/E2PROM - a PROM that can be erased by an electric current. It differs from conventional E2PROM in that it can only be erased in total and not in segments. E2PROMs and Flash were invented by Intel.

LOW POWER SCHOTTKY

A standard series or 'family' of bipolar logic chips usually referred to as 74LS (74 series Low-power Schottky). Invented by Texas Instruments which still dominates the market for them. Various companies now make the same chips but using CMOS process technology instead of bipolar. This makes the chips less power consuming.

MICROPROCESSORS AND MICROCONTROLLERS

A microprocessor is a chip which combines both logic and memory. It forms the central control mechanism in electronic products receiving signals from, and sending signals to, all the other chips and devices in a piece of electronic equipment. Microprocessors were invented by Intel which, with Motorola, dominates the market for microprocessors.

> **A microprocessor is a chip which combines both logic and memory**

Microcontrollers are microprocessors which are used in applications where they only need to be programmed once. They then perform the same function over and over again. Whereas a microprocessor is a general purpose device which is designed to be programmed, say, with Lotus 123 one day and a different spreadsheet the next.

CUSTOM AND SEMI-CUSTOM (generically known as ASICs)

A custom chip is a tailor-made chip which has been built from scratch for a particular customer. As chips are made in layers, or levels, custom chips are sometimes called 'all-level custom' showing that the chip is built from the bottom up completely for one customer's application.

This contrasts with semi-custom where chips are turned out with a number of layers made as standard. The final layers are then customised to fit a particular use.

GATE ARRAY AND STANDARD CELL

These are the main two types of semi-custom chips. Gate arrays are chips containing a number of logic 'gates'. A gate is an arrangement of transistors and capacitors on a chip which form a switch. The 'on' or 'off' of the switch represents the '0' or '1' of the binary code. Gate arrays are chips which have a number of these switches or gates.

To fit a gate array to a customer's particular requirement - say to control the speech of a talking teddy-bear (a very popular application) - it is only necessary to customise one or two 'levels' or 'layers' of the chip. The other layers are standard to all applications.

Standard Cells are semi-customised chips onto which whole-chip functions like RAM, ROM, PROM, EPROM, E2PROM and Microprocessor are placed. In this way a whole electronic system can be designed and implemented on one chip leading to the common expression 'system on a chip'.

PALS, GALS AND PLDS

All these are logic chips which are programmed by the user rather than by the supplier as with 74LS, gate array, standard cell etc. PALs, GALs and PLDs are standard off-the-shelf parts which a user programmes to fit his or her requirements.

PALs, GALs and PLDs are all logic chips which are programmed by the user rather than by the supplier

The differences are:

- a PAL is a one-time programmable bipolar chip whereas GALs and PLDs are reprogrammable CMOS chips.

- a PLD is erasable using ultraviolet but . . . . .

- a GAL is erased using an electric signal.Once erased, both PLDs and GALs can be re-programmed.

WAFERS, DIES AND PACKAGES

Expressions used about chips:

- the wafer is the disc of silicon on which the circuits are manufactured; making the circuits on the wafer ('wafer processing') is the most difficult and expensive part of chip production;

- the die is the individual circuit written on the wafer; after the wafer has been processed it is cut up into individual die;

- the package is an oblong or square piece of ceramic or plastic material with metal legs looking like a beetle into which the die is placed.

The completed chip is then inserted into a:

PRINTED CIRCUIT BOARD

The best way of looking at chips is to regard them as building blocks. It's much the same as a model train set in which the basic building block is the rail. You have straight rails, curved rails, rails which have points in and rails which split into two directions. If you fit all these rails together, you can make a complete circuit around which the model train can run.

If you want to make a piece of electronic equipment, you get the chips and discretes plus batteries perhaps and attach them all to a printed circuit board (PCB). A PCB is a flat board made of resin with holes in, into which the legs of the chips are inserted and then soldered from the bottom side to keep them in place.

The best way of looking at chips is to regard them as building blocks, in much the same way as the basic building block of a model train set is a rail

Before the chips are put in, the PCB will have been printed with lines of metal which will conduct the electricity from one chip to the next so making up a complete circuit.

SURFACE MOUNT TECHNOLOGY

An increasing trend is to have PCBs without holes where the component is stuck with glue onto the surface of the board. This is known as 'surface mount' technology. It is fast taking over from the old technology known as 'through-hole' because machines can stick the components down onto boards at enormous speeds - already at over 100,000 components an hour and increasing.

HYBRID CIRCUIT

In some cases the board will not be made out of resin but out of a ceramic material. In this case the resultant circuit is known as 'hybrid circuit'. Chips and other components are 'surface mounted' onto the ceramic material to make up a circuit. Hybrids are used in cases where the circuitry becomes very hot in use or where especially high reliability or high performance is required.

INTEGRATION

The way the electronics industry evolves is that yesterday's complete system can be put on a single 'board' today (i.e. PCB or hybrid) and will be tomorrow's chip. This is the process known to the industry as ever-increasing 'integration'. It explains why electronics products get continually cheaper and smaller.

The way the electronics industry evolves is that yesterday's complete system can be put today on a single 'board' and will be tomorrow's chip

For instance let's take a computer again. The computers of the Forties were roomfuls of glass vacuum tubes; the computers of the late Fifties and Sixties were collections of discrete transistors on PCBs; the Seventies computers consisted of collections of integrated circuits on PCBs; the Eighties computers, in the case of the PC, are collections of ICs on one board; and already the single chip computer is with us.

To look at it another way, you could say a computer consists of:

- a memory chip to store the operating system (the rules by which the machine works) which are stored in a ROM chip because they will never have to be changed.

- a memory chip to store the work being done that day which is usually a RAM because it allows you to continually cross out (by turning off the power) and then rewrite it.

- a linear chip to take the signals from the outside world e.g. the signals sent from the keyboard as the keys are pressed.

- a chip which converts the linear signals from the linear chip into digital signals which the ROM and RAM can handle.

- some standard logic chips (called 'glue logic') which perform the 'thinking' functions of the computer like mathematics. (Sometimes all the glue logic is put in together on one gate array chip).

- a microprocessor which acts as the central control for all these other chips.

If you take the latest type of each of these six chips you would need to stick them together on a PCB. That would be called a 'single-board' computer.

But, say you took examples of all these six chips from the previous generation of chips and remade them in the latest chip technology - what would happen then?

Quite clearly with each year's advance in the technology allowing you to make the transistors, capacitors etc. half the size they were the previous year (remember 'Moore's Law'?) you could make these six chips so small that they could all sit together on one standard cell chip. That would give you a single-chip computer.

Clearly you don't get the latest in performance when you put all the functions on one chip but you get something small and cheap which doesn't use much power.

SUBSTRATE

A commonly used word in the electronics industry which means the kind of surface onto which you put your circuit. The standard surface is the PCB in which the chips and the metal lines make up the circuit and the resin board is the substrate. If it's a hybrid circuit then the substrate is the ceramic surface; if the circuit is on an IC or chip then the substrate is the silicon on which the circuit has been written.

DISPLAYS

All this talk of chips and substrates, circuits and boards leaves out the part of the electronics product which the user is primarily concerned with - the screen. Basically the chips, discretes and circuitry are all there for one purpose - to put information or pictures up on a screen for the user to enjoy, study or otherwise digest.

> Basically the chips, discretes and circuitry are all there for one purpose - to put information or pictures up on a screen for the user to enjoy, study or otherwise digest

There are four main types of display:

Cathode Ray Tube (CRT)

The commonest type of display which is found in all TVs except the new pocketable sets. They are also those used in all desktop PCs and workstations.

They are shaped like a pyramid and are vacuum tubes with an electron gun at the apex which sprays electrons all over the pyramid's base.

They are power consuming, expensive to make, heavy and bulky but they were the most widely used screen of the Fifties through to the Eighties.

Liquid Crystal Display (LCD)

These are the black and white screens used on watches, calculators and laptop computers. Coloured versions began to appear in the late Eighties for pocket TVs and laptops. LCDs are thin, cheap and use very little power and are therefore very suitable for portable equipment.

Light Emitting Diode (LED)

These are the coloured, flat-screen, bright displays used on instruments, hi-fi equipment and clocks and in cars. They are used where an easily-read, prominent reading is required and where the power source is from the mains or for some other reason does not need to be limited.

Electroluminescent Displays

Very bright, high power consuming displays are suitable for providing information in very large buildings like airports.

A SUMMARY OF COMPONENTRY

These descriptions of the componentry of the electronics industry give a pretty good idea of what the various equipment industries are trying to do: to use electrons running round circuits to store, process, distribute and display information.

How does that definition fit the various industry segments?

- in computers it seems to fit pretty well. Computing, or 'electronic data processing' as it is sometimes called is nothing more than the storing, processing and retrieval of information;

- the definition also encompasses the telecommunications industry which produces equipment which sends speech and data down telephone wires or through the air by radio signals;

- it also encompasses the consumer industry which makes goods which broadcast, receive, store and display 'information' in the form of TV and radio programmes, music, films etc;

- it also encompasses the office equipment industry which makes facsimile machines, copiers, dictation machines, etc;

- it fits the control and instrumentation industry which monitors situations and manufacturing processes and displays the resultant information;

- the medical electronics industry is also covered by the description in that it uses electronics to sense key parameters which indicate a person's state of health;

- the military and avionics industry being largely an amalgam of all these other industries is also covered by the description.

These are the industries which use the products of the electronic components industry. As we have seen from the operation of Moore's Law and the fact that today's board becomes tomorrow's chip, it is the advances in the components technology that allows (you could say obliges) these equipment companies to keep on improving and updating their products.

That means decreasing design cycles, shorter product lifetimes and quicker obsolescence. Does components technology drive the pace of the equipment companies? Or does the pressure of competition on the equipment companies prod them into demanding ever more advanced components? It has been argued about for a long time.

> It Is the advances in the components technology that allows the equipment companies to keep on improving and updating their products

CHAPTER 4

WHERE DOES IT ALL GO?

The sectors of the electronics industry:
computers; consumers; telecommunications;
industrial; transportation; military.

Before the twenty-first century is five years old, electronics will have become the world's largest industry. At the start of the Nineties, it is somewhere in the three quarter of a trillion dollar range - $750bn worth of revenues a year.

> **Before the twenty-first century is five years old, electronics will have become the world's largest industry**

Sometime in the early 2000s it will overtake the other major world industries like car manufacturing, textiles, steel, food, drink, tobacco, chemicals and petroleum to become the largest world industry.

The end of Chapter three gave some idea of what the 'electronics industry' is. Either it's defined as the bit that makes the components or as the industries which make all those goods into which the components go like computers, telephone exchanges etc.

This chapter looks at all those industries into whose products the electronics components go. The idea is to see what these different industries do, what products they make and how technology changes are changing them.

One way to split this $750bn industry is into computers, consumer goods, telecommunications equipment, industrial equipment and transport. If you did that you'd see computers take about $300bn, consumer about $200bn, telecommunications about $110bn, industrial equipment about $110bn and transport about $30bn.

COMPUTERS

The Early Days

First, the fairytale sector of the industry - where the product has shrunk from a roomful to a lapful in 50 years - the computer industry.

> **The computer has shrunk from a roomful to a lapful in 50 years**

Just as the steam engine was a mechanical substitute for horse power, and the Spinning Jenny was a mechanical substitute for muscle power, so the computer is a mechanical substitute for brain power.

The first known example of such a machine was the 5,000 year-old Oriental abacus which allowed you to store numbers as you went along while performing mathematical calculations.

In 1617 the Scotsman John Napier worked out a mechanical way of performing multiplication and division using bone rods - known as 'Napier's bones'.

In 1642, Blaise Pascal of France built a machine which could add and in 1694 Gottfried Leibniz of Germany built a machine that could not only add but also multiply, divide and work out square roots.

In 1835, Charles Babbage of England built a machine which combined all these mathematical calculating capabilities with memory storage. The calculating side could feed back into the memory side and mechanically alter the information stored there.

This interaction between the calculating side of a machine and the memory side is what distinguishes a calculator from a computer. The other key innovation in Babbage's machine was that data was fed into it via punched cards.

In 1859 George Boole of England showed a way to mechanise mathematical logic using the two digits '0' and '1' so providing the basis for binary switching on which modern computing is based. Boole's mechanised logic became known as 'Boolean algebra'.

In 1886, Herman Hollerith of the US improved on Babbage's punched card concept for feeding data into machines by replacing Babbage's mechanical feelers with electro-magnetic sensing devices. In 1911 Hollerith helped form a company called the Computing Tabulating Recording Company which later changed its name to IBM (International Business Machines).

The Modern Age of the Computer

In the early Forties IBM, working with Harvard University, built a machine which could add, subtract, multiply, divide and perform table reference using punched cards. It was 50 foot long and eight foot high.

We've already seen how electronics came to be introduced into the computing scene with ENIAC in 1946. In 1947, John von Neumann worked out a way of building a computer's instructions (its 'program') into the machine itself which allowed it to self-modify its own program in the same way as Babbage's machine had done.

The growth of the computer industry was to be explosive though no-one realised it at the time

From that moment on, the stage was set for a whole new industry to develop. The growth was to be explosive though no-one realised it at the time. Market research commissioned by IBM in the early Fifties forecast a total world requirement of 50 machines. As late as 1956 US market analysts Arthur D. Little forecast total world demand of 600 large and 3,000 medium-sized computers.

The first commercial computers were put on the market in the early Fifties. The purchasers were heavyweight customers such as defence and research establishments. The early leaders in the computer market were IBM of the US and Ferranti of the UK.

While IBM have remained in the lead ever since, Ferranti dropped back to being an also-ran blaming a 'too small' home market in the UK. Its remaining links with the computer business are in specialised military machines. In doing so it missed out not only on the greatest commercial opportunity of the twentieth century but also the most exciting industrial ride of any century. We have seen how the basic building block of electronics equipment went from the vacuum tube to the transistor to the microchip.

By becoming an also-ran, Ferranti missed out not only on the greatest commercial opportunity of the twentieth century but also the most exciting industrial ride of any century

And we have seen how the amount of components you can put on a microchip doubles every year. As might be expected, this progression had a dramatic effect on the size and price of computers.

The Fifties computers were what is called 'mainframes' - machines costing over $1m which required specially trained staff to work them. Often they were leased to customers and the manufacturers would make a steady income from maintenance contracts, providing software and from regular 'updates'.

Since the customers were mainly industrial companies the main manufacturers - like IBM, Univac and Burroughs - cultivated a solid, staid, reliable image. They tended to ape the behaviour patterns of other operations which provide support facilities for manufacturing industry like bankers, accountants, corporate lawyers, stockbrokers etc.

The establishment image of the computer industry could not be sustained, however, in the face of rapid technological change which was always allowing in brash new players. By the end of the Fifties and early Sixties the effects of the transistor and microchip were beginning to show.

A company called Digital Equipment Corporation (DEC) was founded by three men in 1957. One of them, Ken Olsen, is still the boss of a corporation now turning over $12bn.

DEC's first product was called a 'minicomputer'. It reduced price and size by about ten times compared to mainframes. DEC's 1959 machine sold for $125,000; its 1962 machine sold for $27,000; its 1965 machine sold for $18,000 - and each machine delivered more in power than its predecessor. The Great Computer Boom was on.

DEC's first minicomputer reduced the price and size by about ten times compared to mainframes - the Great Computer Boom was on

' The Democratisation of the Computer '

Then, in 1971, came the first microprocessor. It's significance was defined by its inventor, Ted Hoff as 'it allows the democratisation of the computer'. People were not slow to take advantage of the opportunity:

- in 1976 two students Steve Jobs and Steve Wozniak put a computer called Apple on the market which, twelve years later had spawned a $4bn company.

- in 1977 an Auschwitz survivor, Jack Tramiel, launched a computer called the 'Pet' and made himself a multi-millionaire via his company Commodore.

- in 1978-80 Clive Sinclair put his Z80 and Spectrum home computers on the market and made himself a fortune almost overnight.

- in 1980 the founders of another home computer company, Acorn, also made multi-million pound paper fortunes when the BBC adopted their machine as the basis for broadcast lessons in computing.

- in 1981 IBM put a 'microcomputer' on the market which made microcomputers respectable as far as the business world was concerned and a host of companies dashed in to make 'clones' - computers that could run the same software programs as the IBM microcomputer. IBM invented the term 'PC', standing for personal computer which was immediately adopted.

- in 1982 the PC clone company Compaq was started. It made over $100m in its first year and turned into the Nineties on an annual revenue of $1.7bn.

- also in 1982 Sun was founded to make 'workstations' (upmarket PCs for professionals). 'Workstations' differ from 'PCs' merely in their power. Sun was also to beat the $1bn salesmark by the end of the decade.

- in the mid-Eighties, Alan Sugar grew his IBM PC-clone operation Amstrad, into a half-billion pound company.

By the end of the Eighties you could buy the equivalent of a Seventies million dollar mainframe for £500. And still the miniaturisation goes on. The shrinking transistor of the Nineties may make for laptop computers no larger or heavier than a paperback book and costing £50 or so.

CONSUMERS

If you go to an electronics trade show in the West you see men and women in suits talking seriously to other men and women in suits. If you go to one in Japan you feel you're at a party - girls sing the praises of the products, banks of TVs synchronise fast-changing displays, kids play interactive games on the companies' stands.

In consumer electronics Japan is the number one player. One reason why that is so is because when they started their push into exporting electronics goods after 1945, they found the West's military and telecommunications markets under state protection and the computer industry massively dominated by the USA. So there was nothing that was big enough for them to go for except consumer electronics.

Japan now dominates the sector producing over $100bn worth of products a year compared to the $20bn each of the USA and Europe. Non-Japanese Asia accounts for most of the $50bn worth of consumer electronics output of the rest of the world.

Consumer electronics covers: TVs, hi-fi, radios, tape recorders, video recorders (VCRs), cameras, video cameras, compact disc systems (CDs), digital audio tape systems, home computers, car radios, car tape recorders, car CDs, video games, electronic musical instruments like organs and keyboards, watches, clocks and so on.

In many respects consumer is the segment of the electronics industry which drives the development of electronics technology because the consumer is constantly demanding cheaper products than last year or products with more 'features'.

Consumer electronics is the segment which drives the development of electronics technology because the consumer is constantly demanding cheaper and better products

The only way consumer companies can keep up with this is to keep pushing their microchip operations to put more and more features on the chip. This is because the overhead cost of making a chip remains pretty much the same. Therefore the more of the product's features you can put on-chip the fewer chips you use, so reducing overall cost.

Trends in Consumer Electronics

The trends in consumer electronics are towards:

- domestic robots;

- integrated entertainment systems (TV/hi-fi/fax/telephone etc.);

- appliances which respond to spoken commands;

- advanced TV with a lot of memory for storing pictures you want to keep or 'freeze';

- high definition TV (HDTV) which doubles the number of lines used on the screen so making the picture much sharper;

- control mechanisms which announce themselves by speech ('I'm getting too hot'/'Please turn me off' etc.);

- libraries stored on optical discs (CDs);

- interactive learning systems on optical disc;

- interactive games;

- home banking/shopping/travel-booking facilities;

- holographic entertainment and much more.

TELECOMMUNICATIONS

The telecommunications industry is concerned with the transmission of speech and data down wires or through the air. Its principal products are telephone exchanges known as 'public' exchanges, switchboards for offices known as 'private' exchanges, telephones, telex machines, fax machines, and all the associated equipment and gadgetry to connect these to each other.

The telecommunications industry is concerned with the transmission of speech and data down wires or through the air

In the $100bn worldwide telecommunications industry, Japan's output is worth around $40bn, the USA's about $30bn, Europe's about $20bn and the rest of the world about $10bn.

The whole of the telecommunications industry is built on the principle of vibration. Talking vibrates the air around the speaker's mouth. If you put say, a tin near a speaker's mouth, the tin will vibrate; if the tin has a taut string attached to it the string will also vibrate; if someone has an ear against another tin attached to the end of the string, he will hear the transmitted speech.

What Alexander Graham Bell did was to transmit 'vocal or other sounds telegraphically by causing electrical undulations, similar in form to the vibrations of the air accompanying the said vocal or other sounds'. The words come from US patent number 174,465 granted on March 7th 1876 for Bell's invention of the telephone.

The principle is: you speak close to a diaphragm made from a material sensitive to vibration; the speech vibrates the diaphragm; the motion of the vibrating diaphragm produces a corresponding vibration in an electric current; the vibrating current passes along the wire until it reaches an electromagnet; the power of the electrical current fluctuates with the vibration so causing the magnetic force of the electromagnet to vary; if the electromagnet is placed next to a diaphragm made of magnetic material, the varying force of the electromagnet will make the diaphragm vibrate and so reproduce the transmitted speech.

Bell's 1876 patent was the most valuable ever to be issued in the US. Between 1876 and 1893 a number of other claims for patents on the telephone were submitted, and a number of legal attempts were made to annul the Bell patents, but they all failed.

In 1878 the first commercial telephone switchboard came into service - the vast money mine of telecommunications began to be tapped

It was not long before the vast money mine of the telecommunications industry began to be tapped. In January 1878 the first commercial telephone switchboard came into service. It linked up 21 telephones in New Haven, Connecticut, USA.

Two years later the US had 30,000 installed telephones on 138 exchanges. By 1887 the number of subscribers had risen to 150,000 connected on 743 main exchanges and 444 branch exchanges. Despite the British Postmaster General who had called Bell's invention 'daft' the UK had 26,000 subscribers in 1887 - the next largest number after the US. Germany had 22,000; Canada and Sweden 12,000 each; France and Italy 9,000 each, and Russia 7,000.

Government-run Telecommunications

For most of the twentieth century, in most of those countries, the telecommunications industry has been run as a government-controlled operation.

In most countries, telecommunications has been run as a government-controlled operation

Acting through Ministries or 'PTTs' standing for Posts, Telephones and Telegraph, governments have laid down the conditions under which a monopoly supplier can run the telecommunications networks and who is allowed to supply the equipment for it.

For instance in Germany the Bundespost would buy all its exchange equipment from the German company Siemens and anyone wishing to supply telephones or fax machines to run on the system would have to have the equipment approved by the Bundespost.

It was the same pattern in other countries: in Japan the government ran the industry with Nippon Telephone and Telegraph (NTT) running the telephone network; in the USA,

American Telephone and Telegraph (AT&T) ran the network; in Italy it was Italtel; in France - Compagnie Generale Electricite (CGE).

In Britain the same thing happened with the Post Office in sole charge of running the telephone network and deciding who should supply it with equipment. It bought all its exchange equipment from the UK companies GEC, Plessey and STC.

> The telecommunications industry has hidden behind arcane wordage and technical obfuscation to keep its privileges

People have bracketed the telecommunications industry with other successful conspiracies against the layman such as the medieval church or the nineteenth century legal profession. Certainly it has hidden behind arcane wordage and technical obfuscation to keep its privileges.

Different Approaches to Telecommunications

Acting as it does on the borderline between government and industry, the telecommunications industry has spawned many 'fat cats'. But in the USA it could be claimed that AT&T has used its position wisely and responsibly from its dissemination of transistor technology in the Fifties to its backing of the US catch-up programme in chip technology - Sematech - in the Eighties.

Europe was not so lucky. The Post Office controlled who could supply telephones to the system and then only allowed its customers to rent them. It set its own charges and had total control over the technical standards which applied. It was one of the greatest licences to print money devised in the twentieth century, yet the money was not used to promote British technology in the same way as AT&T used its money.

Early in the Eighties the telecommunications side of the post office was split off from the Royal Mail side and BT (British Telecommunications) was born. It was a prelude to rapid change - 'deregulation', competition and privatisation.

The Trend towards Deregulation

In the Eighties, deregulation became the fashion in the US, UK and Japan. AT&T was broken up into separate operating companies in the US, competition was allowed in. The same happened in the UK with Mercury Communications setting up as a competitor to BT. Even in Japan the network was taken out of monopoly control.

The first signs of change for the British consumer was an abundance of telephones in the shops for sale. That was followed by fax machines, answerphones and cordless phones - a range of equipment that was a feast to a consumer who had only been allowed to rent from a very limited range of BT-approved equipment in the past.

With privatisation in the UK and Japan, suddenly the unwieldy, inefficient bureaucratic monsters were having to behave like commercial corporations

The next thing that happened was privatisation in both the UK and Japan with shares in BT and NTT being offered to the public and then quoted on the stock exchange. Suddenly the unwieldy, inefficient bureaucratic monsters were having to behave like commercial corporations.

In Japan and the UK the transition was gentle because there was no initial effective competition. Indeed deregulation allowed BT to get much tougher with its suppliers - for instance ordering Swedish exchanges from Ericsson - while not having to respond to any real outside competition in its own area.

However in the US, AT&T looked very vulnerable because it gave up both a chunk of its network and the right to preferred procurement status on supplying exchange equipment.

Since AT&T has played a key part in developing America's technological prowess, there is genuine concern that a diminution of AT&T's commercial revenues might lead to it having to cut back on its R&D. Any restraint on the inventiveness of AT&T's renowned research arm Bell Labs could have a serious effect on America's technological strength.

Deregulation has brought one benefit to telecommunications - the pace of change is now far faster moving. In the old days the PTTs and monopoly network operators could delay change by taking a long time to establish standards and giving approvals to equipment makers.

The telecommunications industry is at last beginning to move in step with the pace of technological change

Now, with many suppliers chasing the deregulated markets with products of increasing capability, the telecommunications industry is beginning to move in step with the pace of technological change.

Telecommunications Trends of the Nineties

In telecommunications, the trends of the Nineties are:

- further replacement of analogue by digital communication - which will improve quality and aid data transmission

- increased use of mobile, pocketable, digital telephones which could become as cheap and as plentiful as calculators.

- portable fax machines using digital networks.

- transmission of holograms so providing three-dimensional images.

- increasing use of satellites for communication and for navigation e.g. in cars.

INDUSTRIAL

Industrial electronics covers a wide range of areas involved with producing professional electronics equipment and equipment for use in factories. Some of the areas, like control and instrumentation and medical electronics, are regarded as independent sub-sets of the electronics industry in their own right.

The US dominates the sector with output of about $45bn in a $110bn world market. Next comes Japan on about $33bn, Europe on about $27bn and the rest of the world contributing about $5bn.

The industrial sector is very wide, covering: oscilloscopes; process measuring instruments for temperature, flow and level; process control instruments; analytical instruments for analysing materials like spectrographs or x-ray equipment; nucleonic instruments for detecting and measuring radiation; transmitters and receivers for process control systems; signal generators; signalling equipment for road traffic and railways; security systems; fire alarms; echo-sounding or detection instruments; ultrasound instruments; ammeters, voltmeters, ohmmeters and wattmeters; robots; potentiometers, bridges, recorders; aircraft instruments like stall detectors, field detectors and flight recorders.

The big producers in this segment are: Asea/Brown Boveri, Siemens, Hitachi, General Electric of the US, Westinghouse, CGE, Mitsubishi, Cincinnati Milacron, Computervision, Eaton, Fanuc, Foxboro, General Signal, Schlumberger, Fuji Electric, Gould, Hewlett-Packard, Hitachi, Honeywell, Thomson, IBM, Johnson Controls, GEC, Omron, Tateisi and Perkin-Elmer.

Industrial electronics ranges from the simple to the amazingly complex, from a voltmeter to a humanless factory, from a voltmeter to a traffic light system

The segment ranges from the simple to the amazingly complex for example from a voltmeter to a humanless factory, from a voltmeter to a city's traffic light system.

The reason for the segment's diversity is because very often the products have evolved because companies have developed instruments for their own factories and have then seen revenue-raising possibilities for them and have put them on the open market.

Key Industrial Trend

One trend here is towards integrating a total manufacturing process from the design through to manufacture and test (for which the buzz-acronym is CIM ('Computer Integrated Manufacture'). It contemplates developing robots with vision and intelligence; manufacturing and selling humanless factories; and by combining the two to have robots making more robots without human intervention.

'Computer Integrated Manufacture' contemplates developing robots with vision and intelligence; manufacturing and selling humanless factories; and by combining the two to have robots making more robots without human intervention

Within the medical segment you get: X-ray equipment, electrocardiograph machines, pacemakers, diagnostic apparatus, hearing aids, scanners. hospital equipment and monitoring instruments of many kinds. Big companies in this field are: Siemens, a joint GE (of the US) and CGE operation, Philips, Toshiba, GEC, Medtronic and Hitachi.

TRANSPORTATION

This is becoming an area which can justifiably stand in its own right as an industry segment. The big players are General Motors, Ford, Bosch, Nippon Denso, Siemens-Bendix, ITT-Alfred Teves and Magneti Marelli.

The transportation market is potentially explosive as cars include more and more electronic gadgetry and entertainment equipment

It's a market worth round $30bn a year but it can be regarded as potentially explosive as cars include more and more electronic gadgetry and entertainment equipment.

Japan leads the market supplying some $12bn worth of products with the USA second on $10bn, Europe on $6bn and the rest of the world accounting for about $2bn.

Trends are towards safety features like crash avoidance, navigation aids like satellite and CD systems which tell you where you are and how to proceed, and environmental issues like exhaust control.

MILITARY

Many people, when splitting up the electronics industry, would set aside a special segment for military electronics. However, all the other areas are defined by the job the equipment does e.g. computes, communicates etc. whereas military is a hybrid segment consisting of derivative products from all the other product segments.

The main military areas are: radar - either ground, shipborne or airborne; navigational aids - ground, shipborne and airborne; transmitters, receivers and transceivers; encoding and decoding equipment; interference and anti-interference equipment.

The big companies in military electronics are: McDonnell Douglas, General Motors-Hughes, Thomson, Boeing, British Aerospace, Lockheed, MBB-Daimler Benz, United Technologies, Aerospatiale, Martin Marietta, Mitsubishi, Matra, Rockwell, Raytheon, General Electric of the US, Texas Instruments and Harris.

CHAPTER 5

WHAT'S THE CHIP BUSINESS ALL ABOUT? (PART 1)

The importance of chips; size of the market; technology evolution; vast cost of chip research and development and production; erratic chip markets; effect of erratic chip supplies on equipment companies; East/West imbalance in chip production; American approach to chip-making; Japanese approach to chip-making.

The largest and fastest growing part of the components segment of the electronics industry is the semiconductor industry. At the start of the Nineties the semiconductor industry has $55bn+ annual sales and an historical growth rate of 20% per year. It accounts for about 40% of all the total electronics components segment.

As we saw in Chapter three, the semiconductor industry has two branches: the 'integrated circuit' or 'chip' branch, and the other branch called 'discretes'.

THE CHIP
Its Importance, Market and Technology

The difference between a chip and a discrete is that when a single transistor, resistor or capacitor is manufactured and sold as a separate item it is called a 'discrete'. When that transistor/resistor/capacitor is one of many on a piece of silicon it is said to be part of a chip.

Chips are the ultimate vehicle for miniaturisation for the electronics industry

The chip market is growing much faster than the discrete market. That's because chips are the ultimate vehicle for miniaturisation for the electronics industry.

All the billions of dollars, yen, ecus and won poured every year into the development and manufacturing of microelectronics technology are primarily focused on achieving the annual shrink of the transistors, resistors, capacitors and their interconnections which go to make up a chip.

The chip is simply the form in which microelectronics technology is put on the market. Electronics equipment manufacturers can then buy the chip and use it as a building block with which to make electronics products.

The driving force behind the electronics industry is 'Moore's Law' - that you can put twice as many components on a chip every year. That means this year's chip can do the same job as two of last year's chips.

The driving force behind the electronics industry is 'Moore's Law' - that you can put twice as many components on a chip each year

Moore's Law works because no matter how narrow the width of the line along which the electron has to run, and no matter how small the transistor is, the line and the transistor perform their functions as well as a wider line and a larger transistor.

Therefore there is no reason why you can't keep reducing the size of the transistors and the widths of the lines for ever, although technologists have been warning for 30 years that there would be limits. Up to a few years ago they commonly said 'You'll never make line widths narrower than 0.8 microns because that is the wavelength of light'. Now line widths are routinely made less than 0.8 microns.

Presumably we'll come to the end of this particular road when line widths are the same width as the electrons which have to travel down them. But as no-one has been able to measure the width of an electron no-one knows what that line width is.

At the moment, the rule is the same as it's been these past 30 years and looks like being for the next 30 years - that you double the number of features on a chip every year so halving the number of chips you need to make a piece of equipment every year, so halving the size of electronics equipment every year.

The Miniaturisation Process

There are other advantages of the miniaturisation process:

- fewer chips means products cost less (because the cost of making a chip remains constant while you double what it can do every year).

- fewer chips means less power is needed.

- and because less power is needed fewer batteries are needed (and batteries are the main weight problem in portable products like pocket telephones).

Which is why the largest and fastest growing portion of the semiconductor industry is the microchip portion. At the end of the Eighties, the 'discrete' portion accounts for about 20% of the 1989 total of $55bn figure - the monetary worth of the semiconductor industry. But this figure drops every year as the chip takes over.

The Vast Cost of Chip R&D and Production

As well as producing 80% of the revenues of the semiconductor industry, the chip business is also the area which takes most of the money for R&D, factory building and capital equipment.

The sums involved are huge. We saw in Chapter one how the US Congress voted billions of dollars in the Sixties for President Kennedy's moonshot programme much of which found its way into the chip business. That pushed the US far ahead of the rest of the world in microelectronics.

The chip business produces 80% of the revenues of the semiconductor industry and is also the area which takes most of money for R&D, factory building and capital equipment

When the Japanese undertook their big push to catch up with the Americans in chip technology - between 1978 and 1982 - it was estimated by the US Semiconductor Industry Association that the Japanese government and industry invested between $4bn and $5bn over that five year period.

When the European companies Philips and Siemens decided to pursue a catch-up programme in chips in 1984 called 'Megaproject', the cost of the R&D alone was projected at $1bn. In 1988 the cost of a successor R&D project called JESSI (Joint European Sub-Micron Silicon Initiative) is estimated at $4bn.

The Korean programme to catch up in chip technology started in 1983 and has cost around $4bn. The Koreans were still behind the Japanese at the end of the Eighties. The Taiwanese spent $600m on factories and capital equipment in 1989 and that level of spending is likely to continue.

The late Eighties US programme for overtaking the Japanese lead in chip technology - the industry/government consortium Sematech - is expected to cost half a billion a year for as long as it takes to catch up.

And considerably more than the R&D cost is needed if the chips developed in the laboratories are to be manufactured in volume and marketed. The cost of a new commodity chip factory in the early Nineties is around $400m to $500m to set up and costs probably several million a day to operate.

Such factories are obsolete after five years while the selling price of their products declines by 20% per year and every six years there's a major market recession.

So how can the chip industry make profits when it costs such a huge amount of money just to get to the starting gate of competitive technological capability? Especially when the entire world demand is only $45bn?

The answer is you don't make profits at it - or not consistent profits anyway. The industry rule of thumb is that you have to invest $1 to get $1 back in revenues - revenues not profits.

> The industry rule of thumb is that you have to invest $1 to get $1 back in revenues

So why do people do it? Because without microelectronics technology, companies think, they will not be able to compete in any sphere of the electronics industry in the Nineties.

Other electronics companies disagree with that view. They believe that you can just as well buy-in chips and not undergo the expense of making them yourself. But these companies tend not to operate in fast-moving markets where the latest chip technology is a pre-requisite to success such as in consumer electronics.

In consumer electronics it is necessary to continue either producing newer, better versions of products or cheaper products. The only way to achieve that is to put more of the electronics on fewer chips which makes the product cheaper, or to add more features into the existing chips which makes the products able to do more.

So consumer electronics companies reckon it is necessary to have chip technology under their control so they can foresee when the new chip developments will be ready to be incorporated into new equipment products. That way manufacturers can get their equipment products to market ahead of, or at least simultaneously with, their rivals.

Erratic Chip Markets

Companies which have their own chip technology defray the enormous costs of it by selling off general purpose chips to other electronics companies. Companies which do not have their own chip technology have to rely on buying in these chips. But it's a risky strategy for three reasons:

- because without controlled access to the technology a company is unable to foresee accurately when new developments will be forthcoming.

- because there are frequent imbalances of supply and demand in the chip market so prices tend to yo-yo and there are periodic shortages.

- because the biggest chip suppliers are subsidiaries of equipment producers, so the company relying on outside sources of chips is relying on his competitors for essential supplies.

Therefore a company which doesn't have its own chip technology in-house cannot predict either its new product timings or its manufacturing costs. In fast moving markets that is dangerous. Which is why electronics companies without chip operations tend to stay in slow-moving markets like protected national telecommunications and defence markets.

But, as the world moves increasingly away from defence and protected telecommunications markets, electronics companies are looking to the consumer market to replace those lost revenues. And to compete successfully in consumer markets they'll need an advanced chip-making capability.

East/West Imbalance in Chip Production

The governments of the USA and EEC have, in the late Eighties, accepted this argument to the extent that they are giving undisguised subsidies on a massive scale (several billion dollars each) to their local chip industries.

The US and EEC have also tried to protect their local chip industries by trying to fix prices at which chips could be sold. That caused an outcry from US and European equipment producers who said that this price fixing was forcing them out of business.

They could not, they argued, compete in equipment markets with Oriental rivals when the Oriental rivals could buy the chips more cheaply and when many of them had their own in-house supply.

For instance the top Japanese electronics equipment companies NEC, Hitachi, Toshiba, Fujitsu, Mitsubishi, Matsushita, Oki, Sharp, Sony, Sanyo, Seiko-Epson, Rohm and Sanken all have advanced commodity chip-making operations of their own.

But only three out of the top 20 US electronics companies have such capability and in Europe there are only three sources of commodity chips - Philips, SGS-Thomson and Siemens.

The 'Chip Wars'

US and European reliance on Japan's chips had explosive potential and so it turned out in the late Eighties 'Chip Wars' between the US, Japan and Europe. Allegations of dumping, over-pricing, withholding supplies and protectionism were tossed around.

Most dangerous of all was the shortage problem which was a less publicised, but just as potentially a damaging re-run of the Seventies oil crisis. Chips had become, as a prescient US semiconductor boss had predicted, 'the crude oil of the

Eighties'. The Japanese companies were making over half the world's output of chips and, like the OPEC countries of the Seventies, they could hold the world to ransom.

And they did. Major US companies like DEC and Hewlett-Packard publicly stated they could not fulfil their production programmes because they could not get Japanese chips. Amstrad paid $40m to buy 10% of a US chip-maker to ensure supplies.

> **The Japanese companies were making over half the world's output of chips in the Eighties and, like the OPEC countries of the Seventies, they could hold the world to ransom**

HOW DID THE JAPANESE DO IT?

That can only be discovered from looking at the development of the semiconductor industry in the US, Japan, Europe and the Asia/Pacific countries. It's an interesting and revealing tale of national characteristics.

THE SEMICONDUCTOR INDUSTRY AND THE AMERICANS

The problem for all governments was how to overcome the natural conservatism of the business community who always prefer to follow rather than lead when it comes to developing new technologies. They know it can be a costly and often unrewarding business.

> **The governments of all the advanced industrial countries realised that not to compete in electronics was to ensure decline as an industrial nation**

However the governments of all the advanced industrial countries realised that the technology was the key to competitiveness in the electronics industry and not to compete in electronics was to ensure decline as an industrial nation.

The American solution to the problem was typical of the national character. Having reacted to the invention of the transistor by giving away the technology to anyone willing to spend $25,000, the Americans then set out to encourage the companies who were most successful at innovating the technology.

Acting through the US Department of Defence, the leading US computer company IBM, and the leading US telecommunications company AT&T, the American government, throughout the Fifties and Sixties, gave out development contracts to any company willing to pioneer the new technology.

As we have seen in Chapter one a big help to this process in the Sixties was the moonshot programme of President John F. Kennedy which was the magnet which extracted billions of dollars of taxpayers' money from the US Congress.

The effect on the US industrial scene of the policy of financing the most innovative companies was dramatic: the most innovative companies came out on top in the market.

In 1955 the top ten makers of transistors were Hughes, Transitron, Philco, Sylvania, Texas, General Electric of the US, RCA, Westinghouse, Motorola and Clevite. Twenty years later only three of those survived in the 1975 top ten - Motorola, Texas and RCA.

The seven new companies which had sprung from nowhere to make the 1975 top ten were: Fairchild, National, Intel, Signetics, General Instrument, AMI and Rockwell.

Why and How Did the Americans Do It?

The key was the people. Only highly trained scientists, usually with doctorates, could understand and solve the problems involved in developing the technology. The story goes that, in an attempt to impress visiting customers, one of the pioneering companies started using the academic titles of their employees on the intercom. It dropped the practice because 'It sounded like a hospital'.

The snags with employing such people are: they are indifferent to corporate hierarchies and structures; they question every proposition; they won't accept corporate bullshit; they won't take trouble with work they find boring or uninteresting; they are highly egotistical; they can be more interested in solving technological problems than making products for the market.

Such people, however, were the only people who could achieve a lead in the technical race and so get an edge in the market and it soon became apparent that only two kinds of companies could succeed in the chip business: those which were able to manage their technologists well and those which were started and run by technologists.

> It was soon apparent only two kinds of companies could succeed in the chip business: those which were able to manage their technologists well and those which were started and run by technologists

There were an increasing number of the latter kind. The US government's policy of supporting pioneers was to have a prodigious effect on the creation of small companies.

The first one came along in the mid-Fifties when Bill Shockley, one of the three men who had invented the transistor at AT&T's Bell Laboratories, set up a company called Shockley

Laboratories near his home in the area south of San Francisco which is known locally as the 'Bay Area' but which is now known to the world as 'Silicon Valley'.

Shockley put together a fantastic team of people but, although he was a great technologist, he was not the easiest man in the world to work for. In the autumn of 1957 eight of the team walked out on him.

The eight men - Robert Noyce (co-patentor of the first chip), Gordon Moore (of 'Moore's Law), Eugene Kleiner, Jean Hoerni, Vic Grinich, Julius Blank, Jay Last and Sheldon Roberts - are now famous for what they founded in 1957.

They founded both a company and the style of a new industry. The company was Fairchild Semiconductor. The style was a way of working which allowed the scientists, who were pioneering the new technology, to work together in productive groups without such tension-producing elements of traditional industry as hierarchies, status symbols, bureaucracies and politicking.

When Noyce, Moore, Kleiner, Hoerni, Grinich, Blank, Last and Roberts founded Fairchild Semiconductor in 1957 - they founded both a company and the style of a new industry

Noyce took his idea for a new company to Sherman Fairchild who ran a business making cameras and instruments for the military from an office on Long Island, New York.

Noyce persuaded Fairchild that his team could take on the US semiconductor industry's leaders - then Hughes, Transitron, Philco, Sylvania, Texas Instruments, General Electric, Westinghouse, Motorola, RCA and Clevite - and beat them.

The two companies who were the most innovative out of those ten were Texas Instruments of Dallas, Texas, and Motorola of Phoenix, Arizona. Noyce located his new company called Fairchild Semiconductor in the area where the eight former Shockley-ites already lived - California.

In doing so, Noyce sowed the seed for the phenomenon of the 'Fairchildren' - the 150 microchip companies spun-off from Fairchild which were to spawn in Silicon Valley over the next 30 years.

Silicon Valley and the 'Fairchildren' Phenomenon

No-one can explain why the phenomenon of companies spawning new companies happened in Silicon Valley but not in Phoenix or Dallas. One explanation is that Silicon Valley is near San Francisco which is a big international financial centre. That meant venture capital and financial expertise were available to the emerging technologist/entrepreneur breed. Another explanation lies in the entrepreneurial traditions of California from the Gold Rush to the era of the oil wild-catters through to the Hollywood movie days.

Another explanation for the 'Fairchildren' could be the Noyce style of management. Being a scientist working with other scientists, Noyce had a particular style of doing business.

The traditional scientific courtesies allow that every man's opinion has worth irrespective of his position in the company and that every man's contribution is assessed on its content rather than on the job title of the man putting it forward. In other words, a 'consensus' style of management emerged in which every member of the company was involved.

Such a style rules out such practices as: pulling rank, flaunting privilege, politicking, intrigue, ridiculing someone's opinion, and irrationality which are common to the traditional corporate environment.

The Noyce style of management is necessary if complicated science-based products like microchips are to be successfully developed by teams of highly intelligent people, but it also means that these people never get into a corporate way of thinking.

Which could be why, when they have ideas which are different to the current views prevailing in the company - they quit.

Only if you are used to thinking for yourself and having the freedom and confidence to express yourself and get ideas back from others, do you have the confidence to take those ideas to a venture capitalist and set up a company to put those ideas into practice in the real world.

Whatever the reason for the 'Fairchildren' phenomenon, it didn't take long to materialise. In 1959, two years after the foundation of Fairchild, came the first spin-off. It was a company called Rheem Semiconductor. Fairchild sued for appropriation of technology and got $70,000 damages.

By 1960, the three year-old Fairchild was so successful it had made it into the top ten semiconductor companies

In 1960, the three year-old Fairchild was so successful it had made it into the top ten semiconductor companies - at number eight. In 1961 two more spin-offs took place, Hoerni, Last and Roberts formed Amelco (now Teledyne).

In the next six years Hoerni was to found two more semiconductor companies - Intersil and Union Carbide - founding four companies in a decade! Also in 1961 another group formed Signetics (now part of the Dutch group Philips). In 1963 a group left to form General Microelectronics.

In 1965 Fairchild's continuing successes in the field of solid state bipolar logic devices took it into third place in the league of top semiconductor companies behind Texas Instruments and Motorola. Still, the spin-offs went on. In 1966 the operations manager Charlie Sporck went off to run an ailing Connecticut company called National Semiconductor. Sporck turned it into a billion dollar company.

In 1968 the Fairchild board looked outside the company for their third chief executive officer in a year and Noyce quit as general manager and Moore quit as director of R&D. With assistant R&D director Andy Grove they then left to form Intel which became a $2.5bn+ company by the end of the Eighties.

Also in 1968, Jerry Sanders left Fairchild to form Advanced Micro Devices which became a billion dollar company. Others who left over the next decade were John Carey, President since 1982 of the $200m Integrated Device Technology, Jack Gifford who founded the successful linear chip company Maxim Integrated Products and Rodney Smith who founded the programmable logic pioneering company Altera Corporation.

From spin-off companies from Fairchild, other companies spun-off, making a 'family' of over 150 semiconductor companies in Silicon Valley by the end of the Eighties

From these spin-off companies other companies spun-off making a 'family' of over 150 semiconductor companies in Silicon Valley by the end of the Eighties. Undoubtedly the greatest success of all these companies was the one started by Noyce and Moore - Intel.

The Success Story of Intel

Intel was to invent a large part of the standard products which were to form the post-Sixties semiconductor industry. In 1969 it marketed the world's first static RAM - a memory storage device for computers. SRAMs were, at the end of the Eighties, a $2bn annual market.

> Intel invented many of the standard products which were to form the post-Sixties semiconductor industry - and by the end of the Eighties had a $2bn annual market

In 1970, it brought out the world's first dynamic RAM which is also a memory storage device for computers with wider applications than SRAM and, at the end of the Eighties, commands a market worth $4bn a year.

In 1971 it brought out the first EPROM. The EPROM is another memory chip which, unlike DRAM and SRAM, retains its memory when the power is switched off. The 1989 market for it is worth around $2bn.

Another key building block which Intel invented was the E2PROM, an EPROM which could be erased by an electrical signal instead of ultra-violet. The market was emergent in the Eighties and the chip has the potential to take over from DRAM as the biggest selling memory product.

In 1971 the company also invented the world's first microprocessor which at the end of the Eighties commands a $7bn market. Originally designed as a tailor-made chip for a Japanese customer wanting to manufacture calculators, Intel decided that the chip had potential as a general purpose programmable tool. Many equipment companies could make use of the chip by programming it to fit the needs of their particular product.

The concept took off. Instead of having to go to a chip maker to have a chip expensively made especially for their products, equipment companies could buy a programmable chip off the shelf for much less money. It reduced the entry fee for companies wanting to make electronics equipment and it made possible the personal computer boom.

As well as making a lot of money for Intel, the microprocessor gave the company an invaluable lead into the thinking of the world's top electronics equipment manufacturers. That is because a microprocessor is the chip which controls the other chips in a piece of equipment.

That means the choice of the micro has to be made early on in the planning stage of an electronic product. That in turn means that the microprocessor supplier and equipment maker get closely locked in together at an early stage in the designing of electronic equipment.

So the micro supplier gets to know the product plans of many equipment makers which gives him a privileged view of the way the world electronics industry is going. That was a key element in the success of Intel.

The Continued Success of the 'Fairchildren'

Three of the companies which spun off from Fairchild in the Sixties - National Semiconductor, Intel and Advanced Micro Devices (AMD) - became billion dollar companies by around their fifteenth birthdays. Others, like Signetics, became significant companies but under the ownership of other organisations.

Those years were only the beginning of the story of Silicon Valley. In 1978 a capital gains tax break initiated by President Carter sparked off a flood of new companies. In the Eighties, more new microchip companies were founded in Silicon Valley than in the whole of the Valley's previous existence.

Some of these, like Integrated Device Technology, LSI Logic, Cypress Semiconductor and Chips and Technologies are $200m+ companies at the end of the Eighties and striking out to take over from the start-ups of the Sixties who in their turn took over from the industry leaders of the Fifties. That was the American way of doing it - entrepreneurial, venture capital-backed, classic capitalism.

THE SEMICONDUCTOR INDUSTRY AND THE JAPANESE

The Government of Japan in the late Fifties and Sixties determined to become a world-class player in the computer industry. In order to achieve that goal, the government decided that it needed to be world-class in the most important ingredient of computers - chips.

The Japanese Government was determined to be a world-class player in the computer industry and took strident steps to get there

Government Steps

Acting through the Ministry for International Trade and Industry, the Japanese Government went about achieving this goal in the following ways:

• it put steep tariffs and restrictive quotas on imports of advanced microchips;

• it required US microchip firms seeking access to the Japanese market to license their technology to local firms;

• it insisted that local firms could only license US technology if they could utilise it to produce microchips for export and if they would spread the technology to other Japanese companies through sub-licences.

The extent to which this policy was implemented can be seen from the fact that in 1969, Japanese microchip producers were paying royalties to the US that were equal to 10% of their sales - mostly to Fairchild and Texas Instruments.

This may have given the Japanese companies the ability to copy US microchip technology but it did not give them the ability to develop their own. Nor were they trying to - at the beginning of the Seventies the combined spending of the three largest Japanese producers - NEC, Hitachi and Fujitsu - was less than the R&D budget of the leading US microchip producer Texas Instruments. So, in 1971, the government of Japan embarked on a policy designed to boost private sector R&D investment in microchips. MITI encouraged the pairing of the six major companies into three groups - Fujitsu-Hitachi, NEC-Toshiba and Mitsubishi-Oki to promote specialisation of developments. Each group received government subsidies worth around $200m between 1972 and 1976.

One measure of how advanced the production of a semiconductor company was at that time, was the proportion of its discrete components sales to the proportion of its chip sales.

In 1968 the proportion of chips was worth only 10% of the total Japanese semiconductor production figure of $250m. In 1978, that proportion had risen to 50% on a total semiconductor production figure of $2.4bn.

In 1975, MITI embarked on its VLSI (Very Large Scale Integration) programme which had the aim of technical parity with the Americans by the early Eighties. The VLSI programme cost about $65m a year over five years.

Nine years later, it was clear that the money had been well spent. In December 1983, NEC began mass production of 256K DRAMs about 12 months before similar levels of production were reached by any American manufacturer.

Japanese Dominance

That signalled Japanese dominance of the industry. By the end of the 1980s Japan was outproducing America in semiconductors (chips and discretes) and was making over half the world's total production. It was also accounting for over half the world's usage.

> By the end of the Eighties, Japan was outproducing America in semiconductors and was making over half the world's total production

A major contributor to the Japanese success was the industry's access to large amounts of capital at low interest as a result of the practice of major Japanese companies to form themselves into conglomerates grouped around a major bank.

This provides great strengths to members of the conglomerate going through difficult trading conditions. This is buttressed by the highly diversified nature of Japanese electronics who compete in many segments of the electronics industry. This helps them to subsidise the notoriously severe downturns in the microchip business with profits earned elsewhere.

A further advantage of this close industry/banking relationship allowed industrial companies to borrow heavily for their capital needs. Many semiconductor firms in Japan have debt-to-capital ratios in the 60% to 70% range compared to the sub-20% ratios of US companies.

With astute management by the Government of Japan through MITI, and a supportive industrial structure, the Japanese chip business appears to be on course to take over the world. At the end of the Eighties Japan is making half the world's output of semiconductors and about 45% of the world's output of ICs.

With astute management by the Government of Japan through MITI, and a supportive industrial structure, the Japanese chip business appears to be on course to take over the world

The 1988 world semiconductor league table was headed by three Japanese companies: NEC, Toshiba and Hitachi; in fourth and fifth places were two US companies: Texas and Motorola; in sixth, eighth and ninth place were the Japanese companies Fujitsu, Mitsubishi and Matsushita; Intel of the US was seventh and Philips, the only European company in the top ten was tenth.

THE CHANGES IN THE STRUCTURE OF THE CHIP INDUSTRY

The Changes in the Top Ten World Companies

The changes in the structure of the chip industry in the period of the Fifties to the Eighties can be seen from the changes in the list of the world's top ten companies.

Between 1955-75:

- Hughes, Transitron, Philco, Sylvania, General Electric of the US, Westinghouse and Clevite - all from the US

- were replaced by Fairchild, Intel, Signetics, General Instrument, AMI, Rockwell and National - all from the US.

Between 1975-88:

- RCA, Fairchild, General Instrument, AMI, Rockwell and National - all from the US

- were replaced by NEC, Toshiba, Hitachi, Fujitsu, Mitsubishi and Matsushita - all from Japan.

The message for the Americans, who had invented the chip industry and regarded it as their divine right to own it, was traumatic. If this rate of progress of Japan's chip industry was to continue unchecked, the prospects for America in the chip business were dismal.

Furthermore, America's consequent dependency on Japanese chips would undoubtedly result in Japan achieving what its government targeted back in the Fifties - supremacy in the world computer industry.

CHAPTER 6

WHAT'S THE CHIP BUSINESS ALL ABOUT? (PART 2)

Different approaches to chip-making in the countries of Europe and Asia.

THE SEMICONDUCTOR INDUSTRY AND THE EUROPEANS

Megaproject and JESSI are by no means the first attempts by Europe to produce successful government-backed initiatives in microelectronics - far from it.

From the first emergence of the chip business in America in the early Sixties the governments of Europe recognised that this was going to be the most important industrial process technology of the second half of the twentieth century.

European Governments recognised that the semiconductor industry was going to be the most important industrial process technology of the second half of the twentieth century

As in America, the biggest hurdle to the governments of Europe in ensuring the technology was mastered, was the conservatism and risk-aversion of the established, traditional electronics companies.

Over the years the big electronics companies of Europe have demonstrated a common approach to the emerging microelectronics technology - they decided that the best way of getting access to it was either to buy up American or Japanese companies which had it, or to buy licences from American and Japanese companies which would sell their technology.

For instance Siemens bought the American companies Litronix, Databit, Threshold Technology and Microwave Semiconductor and a 20% stake in Advanced Micro Devices in 1977. It also set up a joint venture with Fuji Electric of Japan and licensed the right to make Intel's microprocessors, to make Rockwell's bubble memories and to use Toshiba's CMOS process technology.

Philips bought the US chip firm Signetics in 1975 which has given it a channel into the thinking of Silicon Valley ever since. And in 1953 took a 35% stake in the active components activities of Japan's Matsushita Electrical Industries. For many years it has had technology deals with Intel.

The Italian semiconductor company Societa Generale Semiconduttori (SGS) was one-third owned by the Californian company Fairchild until 1968 and later made technology exchange deals with Toshiba of Japan, and the US companies Zilog and LSI Logic.

In the UK Ferranti bought Interdesign of the US and Lucas took a 24% stake in the US firm Siliconix. In the Sixties, Marconi (later owned by GEC) bought a licence from Fairchild to make early logic chips in the UK, and in the late Seventies GEC again went to Fairchild for technology intended to make memory chips - a project that was later abandoned.

In France, the country's largest components manufacturer, Thomson, licensed technology from Motorola of the US while the French defence manufacturer, Matra, bought chip technology from the American companies Harris, Intel and Cypress Semiconductor. Another American company, National Semiconductor, provided chip technology for a joint venture with the French company St. Gobain.

European Governments' Intervention

Clearly the big European companies were of one mind in thinking that the best way of controlling access to the technology was to buy it in rather than develop it themselves. However, the governments of Europe had other ideas.

Believing that their indigenous computer industries could never succeed without home-grown chip technology and that their national security would be under threat if they depended on foreigners for microchips, the various national governments of Europe sought to cajole, push and manipulate their indigenous electronics companies into mastering microelectronics technology and competing in the world's chip markets.

Despite all the efforts of governments, Europe remained lagging behind the Americans and Japanese in the commercial microelectronics technology field

But throughout the Sixties, Seventies and Eighties, despite all the efforts of governments, Europe remained lagging behind the Americans and Japanese in the commercial microelectronics technology field.

One response of European governments to this technological lag was to maintain a high tariff (initially 17% later 14%) on chips imported from America and Japan. One effect of the tariff was to encourage producers to set up manufacturing locations in Europe and in this respect it was very successful. Many did: Texas Instruments, National Semiconductor and Motorola of the US all put in factories in Europe and so did NEC, Hitachi, Toshiba and Fujitsu of Japan.

But the longer term aims of the European governments in promoting this inflow of foreign companies were not fulfilled. The governments had hoped that from these foreign companies technology would disseminate into the local industry through a repeat of the 'Fairchildren' phenomenon seen in California.

It did not happen. The conditions for entrepreneurial activity in semiconductors did not appear to exist in Europe, or did not exist to the same extent as they did in California. Europeans who wanted to be engaged in new chip company start-ups commonly went to California to do it, and some did it very successfully but, except for some heavily government-subsidised and none too successful operations, start-up activity did not materialise in Europe.

So European national governments looked for more direct means to prod their local industries to invest in the R&D for chips, to set up new chip factories and to compete with the Americans and Japanese in world chip markets. From country to country the objectives were the same but the means the various governments adopted to achieve them were very different.

GERMANY

The German government has one very simple way of subsidising its national chip business - it gives money to Siemens. As the country's biggest electrical/electronics company with a turnover at the end of the Eighties of around $30bn, Siemens, has always been by far the largest producer of microchips in Germany.

> **The German Government has a simple way of subsidising its national chip business - it gives money to Siemens**

Between 1970 and 1980 the government of West Germany contributed one fifth of the $200m Siemens spent trying to close the technology gap with the US and Japan.

It wasn't very well spent. In late 1981, when the company's 64K DRAM was put on the market, Siemens was four years behind the market leader. It should be noted, however, that Siemens was the first European company to market a 64K DRAM.

In 1983 the German government in conjunction with the Dutch government decided to back Philips and Siemens to continue the catching up process. That programme was called the Megaproject.

The German government is reported to have been furious when Siemens bought-in technology from the Japanese company Toshiba to enable the German company to manufacture the first product required under the Megaproject - a 1 M-bit(1 Megabit) DRAM.

Five years and $1bn on from 1983, when Siemens' 1 M-bit DRAM appeared on the market the company was eighteen months behind the market leader which was Toshiba.

By the end of the Eighties, Siemens reckoned it was only six months behind the market leader with the second product of the Megaproject - the 4M-bit DRAM.

FRANCE

In France the government's efforts to close the technology gap have been considerably more diverse. Until 1978 there only was one French chip firm - Thomson - and the French government encouraged it to become a broadline supplier of commodity chips.

The French Government's efforts to close the technology gap have been very diverse

As such, Thomson's chip business was chronically unprofitable and since the mid-Sixties the French government has pursued a policy of subsidising chip R&D and offering Thomson preferential purchasing arrangements with protected markets for military and telecommunications chips.

For instance the big French telecoms company Alcatel, a subsidary of Compagnie Generale d'Electricite (CGE), was obliged to buy much of its chip requirement from Thomson. The same applied to defence manufacturers.

So much so, in fact, that Thomson's chip production capabilities became so exclusively tilted to military and telecoms equipment that in 1968 the French government gave the company $18m over five years just to learn how to produce chips for computers.

Thomson also had to resort to government-subsidised exports to Eastern Europe and less developed countries where competition from the Americans and Japanese could be largely avoided.

Government Plans

Then came the French government's 'Integrated Circuits Plan' of 1978 which offered government support for leading edge commercial production of standard chips. In order to qualify for the grants, companies had to show that it had access to the technology. This led to the founding of two joint venture companies with American chip firms.

First, in 1978, Matra formed a joint venture company with Harris of America known as Matra Harris Semiconducteurs; second, in 1979, St. Gobain formed a joint venture company with National Semiconductor known as Eurotechnique.

A third result of the government initiative was a technology assistance agreement negotiated by Thomson with Motorola of the US. Later on, Intel joined in the Matra-Harris joint venture and, later still, Matra Harris went to a third US firm, the 1983 start-up company Cypress Semiconductor for more process technology.

In 1981, came the French Government's second 'Integrated Circuits Plan' which directed money specifically at the most advanced microelectronics products and processes. In 1982 the government directed $70m into chip R&D and over the next four years a further $500m was funnelled into this purpose.

At the same time the IC Plan Mark 2 gave some $40m to double the size of the government-owned Centre National d'Etudes Telecommunications (CNET) and to buy MOS technology from the American firm National Semiconductor.

The effect of the French government's prodigal dispensation of chip assistance was to cause a surfeit of production capacity in France and consequent financial problems for all the producers.

In 1984 National pulled out of Eurotechnique leaving the French government to absorb the joint venture into Thomson. In 1985 Thomson bought the US chip firm Mostek for the knockdown price of $70m. Mostek's parent company United Technologies Corporation (UTC) had bought Mostek for $380m in 1979 and it had cost UTC over $1bn to keep Mostek afloat until it sold it off to Thomson. So Mostek cost UTC $1.3bn in five years.

Another ruinously costly intervention in the chip business occurred in 1979 when the French oilfield services company Schlumberger bought Fairchild for $425m. Over the next eight years Schlumberger was to invest a further $1bn in the chip company. In 1987 Schlumberger sold Fairchild on to National Semiconductor for $122m - representing a loss of $1.3bn in eight years.

In 1988 the French and Italian governments decided to merge Thomson and the Italian semiconductor company SGS. Throughout the decade Harris became increasingly disenchanted with Matra Harris Semiconducteurs and gradually reduced its equity stake finally giving up all financial involvement in the venture in 1989.

BRITAIN

Like the French, German and Japanese governments, the British government believed that if it was to have an independent computer industry it needed an indigenous chip industry. Consequently it has supported the chip business in various forms throughout the last three decades.

The British government believed that if it was to have an independent computer industry it needed an indigenous chip industry

In 1967 the UK government granted its first two contracts for upgrading semiconductor production processes. One went to Elliott Automation to install an MOS process (the year before Intel was founded explicitly to develop MOS), and another to Ferranti.

In 1968, UK government R&D for semiconductors amounted to $8m which was a third as much as the USA was spending and more productive bearing in mind the cheaper research costs of the UK.

Between 1969 and 1973, the government, through the National Research Development Corporation, channelled R&D and production refinement funds into Marconi-Elliott Microelectronics, Plessey and Ferranti.

However the late Sixties saw a government sponsored reorganisation of the electronics industry with the government providing funds to one merged company comprising GEC, AEI and English Electric.

The Reorganisation

The reorganisation preceded a series of cut-backs in the UK manufacturing of semiconductors. Following a 1968 price-cutting move by the UK's largest supplier, Texas Instruments, GEC sold off most of its shares in its joint semiconductor manufacturing venture with Philips and Associated Semiconductor Manufacturers (ASM). ASM was, at the time, the UK's second largest semiconductor manufacturer.

In July 1971, GEC closed down the Glenrothes microchip factory of Elliott Automation and the Witham chip factory of Marconi-Elliott Microelectronics so putting the UK out of the market for standard microchips.

The other main UK companies in the chip business, Plessey and Ferranti, were both in custom or niche markets which were too small to adequately finance R&D without continued external assistance.

In 1978, GEC attempted a come-back in the market for standard microchips through a government-assisted joint venture with Fairchild called GEC-Fairchild which had intended to make standard memory and microprocessor chips. Following the takeover of Fairchild by Schlumberger, the venture was abandoned.

In 1978 the government tried another initiative to get the government into standard microchips by backing the start-up company, Inmos. The financing was arranged on the model of a Silicon Valley start-up and the products were standard memories and microprocessors.

The initial 1978 funding was $93m. In 1980 the British Technology Group invested a further $60m. In 1983 it invested a further $23m. That year it was sold to Thorn EMI for $124m who held it until 1989 when it handed the company on to SGS as a virtual gift. Inmos cost Thorn some 300m in all.

A further 1978 UK government initiative was the establishment of a five year $200m initiative to subsidise collaborative microelectronics R&D principally GEC, Plessey and Ferranti. GEC was later criticised for pulling out of the CMOS projects earlier than its partners.

Another UK initiative to get back into standard microchips came in 1983 when STC built a large factory with the intention of making DRAMs and SRAMs. The company ran out of money before it could install the processing equipment.

In 1987 Plessey bought Ferranti's semiconductor division and merged it into Plessey Semiconductor. In 1989 a takeover bid for Plessey by GEC and Siemens left Plessey Semiconductor 50/50 owned by GEC and Siemens and 100% managed by Siemens with an embargo slapped on it by the EEC to the effect that it had to be run as a separate company.

At the end of the Eighties, the UK had half a $300m chip enterprise, and two chip businesses each worth around $20m - it was not much to show for three decades of government support

That left the UK owning, at the end of the Eighties, half a $300m chip enterprise (Plessey), and two chip businesses each worth around the $20m mark (GEC and STC). It was not much to show for three decades of government support.

HOLLAND

Philips

Semiconductors in Holland means Philips. Philips is the largest producer in Europe, tenth largest in the world, and the largest supplier to the European market.

Philips has a long record of internal innovation in the semiconductor business, for instance the germanium alloy junction transistor, the Gunn diode, the diffused transistor, integrated Schottky logic and voice synthesis chips.

Semiconductors in Holland means Philips - the largest producer in Europe

Just as in Germany with Siemens, if the government of Holland wants to give money to support the electronics industry, it gives its money to Philips. And, as in Germany with Siemens, it doesn't always get what it thinks it has paid for.

For instance, in 1983, the Dutch government announced assistance for the Philips/Siemens 'Megaproject' to catch up the Japanese in RAMs. The Dutch government is thought to have been disappointed that, by the end of the Eighties, Philips had still not gone into production with the SRAMs developed as part of the Megaproject.

Nonetheless the Dutch government will be contributing a quarter of the costs attributable to Philips of the Megaproject's successor programme JESSI. However the EEC, which is contributing a quarter of the total cost of JESSI is thought to have insisted on some very tight procedures to force participating companies to commercialise the products developed under its auspices.

Why Many New Chips Are Not Sold On The World Market

It might be thought odd that companies do not want to recover some of the huge costs incurred in developing these products by selling product in the marketplace.

The reason why companies develop new chips but don't market them is because the mass manufacturing and global marketing of chips is an expensive and risky business because many other companies are also trying to defray their R&D costs by selling chips for anything they can get. This makes it a cut-throat, unstable, often low-price market.

Companies develop new chips but don't market them because the mass manufacturing and global marketing of chips is an expensive and rIsky business

Therefore where governments help out with development costs, companies often do the development and then keep the technology for in-house needs - which are basically adding value to their equipment products.

However if the companies take this line, the governments don't get what they wanted which is national chip industries capable of supporting themselves without continual re-infusions of taxpayers' money.

National governments also want to provide for 'import substitution' i.e. keep a balanced trade in microchips, to resist dependency, to employ its clever graduates rather than see them go abroad, to spread the technology in their countries and promote a lively, competitive internal chip industry capable of supporting the local equipment industry and local defence needs.

This does not happen if companies take government money, develop the technology then keep it all in-house.

ITALY

As in Holland, Italy has only a single semiconductor company, SGS (Societa Generale Semiconduttori) which gets all the government money handed out to promote commercial microelectronics. Since the merger with Thomson of France, SGS is a Franco-Italian company and so also gets money from the French government.

SGS' Italian parent is STET which is the Italian government-owned telecommunications company. Its French parent is the Thomson holding company. In 1989 it acquired the UK government-funded start-up company Inmos. It has technology links with Toshiba of Japan and Zilog the US microprocessor company and LSI Logic the US gate array company.

Under the ex-Motorolan Pasquale Pistorio SGS has committed itself to becoming a global supplier of semiconductors in the Nineties. It is moving into all the commodity chip areas - memory, logic and microprocessors - and is manufacturing in Asia and the US as well as Europe.

It is also involved in JESSI with Siemens and Philips and this programme should help it keep up with the underlying technology without straining the financial resources of the SGS' and Thomson's State backers.

The backers, so far, seem committed to supporting the company for as long as it takes to get it established as a world top ten chip company capable of supplying European electronics equipment companies with their commodity chip needs. The only cloud over SGS-Thomson is that governments are traditionally fickle backers of anything.

SGS-Thomson seem to have backers who are prepared to support the company for as long as it takes to get it established as a world top ten chip company

A SUMMARY OF EUROPE'S INVOLVEMENT IN THE CHIP INDUSTRY

A summary of the results of 30 years of European attempts to succeed in the chip business:

- At the end of the Eighties, Europe contributed about $5bn to a $55bn world industry.

- The combined sales of the top five European companies were about $4.3bn whereas those of the top five Japanese companies were about $17.3bn.

- Europe had the world number 10, Philips with sales around $1.8bn; the world number 14, SGS, with sales around $1bn; the world number 20, Siemens, with sales around $800m; the world number 31, Telefunken with sales around $300m; and the world number 32, Plessey, with sales around $280m.

- Europe also had some minor companies: Asea (Hafo) and Ericsson (Rifa) both of Sweden, Matra-Harris, European Silicon Structures, Marconi, STC, Mietec of Belgium (a subsidiary of Alcatel), Austria Micro Systems and a host of purely design companies.

THE SEMICONDUCTOR INDUSTRY AND ASIA/PACIFIC

KOREA

In 1983 Korean chip production was worth half a million dollars. By the end of the Eighties it was worth one and a half billion dollars. How did it happen?

Simple, $4bn. Between 1983 and 1989, that sum was spent investing in chip R&D and new chip factories by Korea's Big Three in chips - the conglomerates ('chaebols' to give them their Korean name): Samsung, Hyundai and Lucky Goldstar.

The country has two smaller chip operations, Daewoo and the Korean Electronics Company, but the Big Three account for 98% of the Korean industry's chip output. Samsung is much the biggest of the Big Three topping the $1bn mark for chip production at the end of the Eighties. The other two are about a quarter Samsung's size in chips.

Korea's semiconductor industry became the world's third largest after America and Japan as early as 1985

Taking semiconductors overall - i.e. chips plus 'discretes' - Korea's semiconductor industry became the world's third largest after America and Japan as early as 1985.

It was a big change from its status as a cheap labour area for US companies. Starting with Motorola, Fairchild, Signetics and AMI, many US companies set up chip assembly plants in the country in the Seventies. The country has the world's largest chip assembly operation - Anam, an affiliate of the US company Amkor.

The Japanese followed suit. The Korean Electronics Company was formed in 1969 as a joint venture with Toshiba. Other Japanese companies came in during the Seventies - Sanken, Sanyo, Rohm, Mitsumi, Toko.

As well as Anam, a number of independent Korean sub-contract semiconductor assembly companies like Integrated Circuits Packaging, KTK and New Korea Electronics were set up.

In the early Eighties, however, the Korean government started to take an interest in the wafer processing side of the semiconductor industry. The government had realised that despite all the US and Japanese plants in the country, and the sub-contract assembly houses, the local Korean electronics equipment industry was not getting the chips it needed from local sources.

Instead of producing semiconductors for the local electronics industry, the foreign-owned plants were producing chips for export. Consequently, the microchip requirements of the Korean electronics equipment industry had to be almost totally imported.

Since 1979, the government of Korea has tried to improve the technological capability of local companies. The government set up the Korean Institute for Electronics Technology (KIET) with $60m in backing which specifically targeted chip production technology.

KIET set up prototype design and production facilities in Korea. Hewlett-Packard and AMI of the US provided help in the training of KIET people in process technology and circuit design. RCA provided testing know-how; Fairchild provided software expertise. A KIET office in Silicon Valley tapped the locality's technological capabilities.

KIET makes available to Korean companies the technology it acquires. It does this both directly and indirectly: by licensing the technology directly in return for fees, and by allowing ready access between industry personnel and KIET staff.

For instance, Samsung's first 8-bit single chip microcomputer was developed by KIET at a cost of $800,000. Samsung paid KIET one-third of the development cost for the rights to sell the chip.

The Success of Korea

In 1982, the Korean government announced the 'Semiconductor Industry Promotion Plan'. The aim of the plan was first to provide for import substitution; later on to provide chip exports. As we have seen, it worked brilliantly. The boom year for the semiconductor market of 1988 saw a remarkable surge in the Korean companies' chip (IC only, not discrete) output: Samsung up 192% in a year; Goldstar up 103%; Hyundai up 253%.

The Korean government's 'Semiconductor Industry Promotion Plan' worked brilliantly - the boom year for the semiconductor market of 1988 saw a remarkable surge in the Korean companies' chip

Samsung had a remarkable success in memory chips - principally DRAMs - and though it had still not caught up with the Japanese in DRAM technology by the end of the Eighties, it was probably not much more than a year behind at the 4 Megabit stage.

The Koreans did not do it all from home-grown capabilities. Samsung licensed technology from Micron Technology, Exel and Intel of the US; Lucky Goldstar's semiconductor operation, which was from 1981 a joint venture with AT&T of the US, acquired technology from AT&T, Zilog, LSI Logic and AMD of the US; Hyundai licensed technology from Inmos of the UK and the Western Design Centre in the US.

Hyundai also tapped into the talent of a Silicon Valley subsidiary called Modern Electrosystems to access technology, while Daewoo did a technology exchange deal with the Canadian telecommunications company Northern Telecom.

The Korean companies also gained technology from acting as foundries for American companies. For instance Hyundai made DRAMS for Texas Instruments, E2PROMs for Atmel and acted as a manufacturing source for many small US companies without a manufacturing capability of their own.

Goldstar had a supply deal from 1984 to 1986 with LSI Logic of the US to supply it with gate array wafers. Samsung has supplied DRAMs and EPROMs to Intel and has acted as a foundry source for various US and Japanese manufacturers.

Funding for Success

The extent of the funding provided by the government of Korea and the World Bank in achieving this turn-round in the Korean semiconductor industry's capabilities cannot be accurately assessed. It must have been a fairly open-ended commitment because Samsung's technology catch-up strategy included simultaneously building three chip factories - each for different DRAM generations (64K, 256K and 1M-bit).

The normal practice is to fund the building of factories to produce one generation of product with profits generated from selling the last generation of product. That implies there were none of the normal commercial pressures on Samsung. Without huge outside support, the Samsung strategy would have crippled the largest industrial companies in the world.

Samsung's gung-ho entry to producing three generations of chip simultaneously astonished the world semiconductor industry. No-one had ever tackled the chip business with such open-handed determination before

So Samsung's gung-ho entry to produce three generations of chip simultaneously astonished the world semiconductor industry. No-one's ever tackled the chip business with quite such open-handed determination before.

TAIWAN

As with Korea, Taiwan's earliest introduction to the semiconductor industry was from foreign companies setting up assembly facilities. Those that did so were General Instrument, Siliconix, and International Rectifier of the US, Mitsubishi of Japan and Philips of Holland.

Like Korea, Taiwan's earliest introduction to the semiconductor industry was from foreign companies setting up assembly facilities

Taiwan also has a number of indigenous semiconductor assembly operations such as Farsonics, Fine Products Microelectronics Corp, Oak East Industries, Orient Semiconductor Electronics and Unitron Industries.

The country's first moves into wafer processing came in 1976 when the Taiwanese government's research body, the Industrial Research Institute (ITRI) purchased CMOS process technology from RCA. They transferred it into its own laboratories and began a continuous programme of R&D on the process, aiming to improve it.

At that time the dominant world process technology was NMOS, but the Taiwanese wanted CMOS because it made chips which used very little power and so were suitable for the products of the Taiwanese electronics industry which at that time was heavily into manufacturing calculators, watches and telephones which require very low-power usage.

ITRI's electronics arm is a body called Electronic Research Service Organisation (ERSO). In the late Seventies, ERSO took a 10% share in Taiwan's largest semiconductor company, United Microelectronics Corporation (UMC), in exchange for transferring to UMC its CMOS process technology. The Taiwanese government provided financial assistance for factory building and, between 1980 and 1984, doubled its annual expenditures on chip R&D to over $90m with a rising commitment to reach $325m a year by the end of the decade.

In early 1982, UMC opened the country's first commercial chip factory - a $21m plant aimed at supplying the Taiwanese and Hong Kong markets with chips for calculators, toys, watches and telephones.

UMC was ably and shrewdly managed from the start, building its business on supplying specific professional markets and eschewing the mass market for commodity memories. That way UMC was able to maintain its profitability - no easy thing to do in the chip business. In 1988 it had $22m profit on sales of $108m.

In 1983 Taiwan introduced a new concept into the chip business based on the Silicon Valley 'start-up' model. Three companies called Vitelic, Quasel and Mosel were founded to pioneer commodity memory technology developing DRAMs and SRAMs.

Because the trio used Taiwan government money, US investment money, US design talent, Taiwan factories and sold into US markets they were dubbed 'shore-to-shore' operations.

Vitelic was the only one of the three to have done anything significant in the marketplace by the end of the decade when it was selling around $45m worth of speciality and high speed DRAMs a year.

Further Government Initiatives

In 1984 came another government initiative. Disappointed by the failure of foreign companies to put chip plants in Taiwan, despite the blandishments offered in the shape of financial assistance to set up in the Hsinchu Science Park (modelled on California's Stanford Industrial Park), the government proposed a chip foundry.

The proposal was that in return for putting up their money, foreign companies could have a share of a chip factory which would have no products of its own but which would make its money by processing the designs of other companies.

Set up under the chairmanship of Morris Chang, ex-CEO of General Instrument, the venture became known as the Taiwan Semiconductor Manufacturing Company (TSMC). Philips and the Taiwan government were the major investors and the plant's capacity was quickly reached. It has been expanding its capacity steadily ever since.

The venture appears to be a success on three fronts: as a commercial operation in its own right; in attracting a host of new design companies to set up in Taiwan with the aim of using TSMC's factory capacity; and in making innovative design and the latest process technology accessible to the Taiwan electronics equipment industry. The latter has grown by leaps and bounds in the last few years.

The government's pump priming policy seems to have been successful in Taiwan. The first chip companies to be set up without any government funding - Hualon Microelectronics Corporation, Winbond Electronics Corporation and Advanced Microelectronic Products - have combined 1989 sales of $100m. Hualon and Winbond are spending $360m on new chip-manufacturing plant.

Total Taiwanese chip production was worth $250m in 1989 and total investment in plant and capital equipment was $600m. UMC's new $100m factory started production in 1989 and a new one is already planned - a $270m affair using the new eight inch diameter wafers. These allow you to double the amount of die you get from a wafer without significantly increasing the cost of producing a wafer.

The Taiwan government has been more successful than any government except Japan's in producing a thriving chip industry. An industry which is able to attract international investors, to produce profits, to become self-supporting and even to spawn new commercial ventures without government intervention.

The Taiwan Government has been more successful than any government except Japan's in producing a thriving chip industry

The island is now projecting a $3.5bn chip industry by the end of the Nineties.

HONG KONG

Hong Kong has the distinction (if it can be called that) of hosting the first ever off-shore semiconductor assembly operation with the establishment of Fairchild's operation in 1962.

Others followed: National, Siliconix, Teledyne and Silicon General of the US; Hitachi and Sanyo of Japan. In addition Hong Kong has indigenous assembly operations: Semiconductor Devices Ltd - a subsidiary of Wheelock Marden and the biggest assembly operation in Hong Kong; Century Electronics; Elcap; Micro Electronics, and RCL.

Three of these moved into wafer fabrication during the Eighties starting with Elcap in 1982, and later RCL and Hua Ko.

Two of these are ultimately controlled from Peking and are thought to be useful in allowing China to get access to chip technology. One of these is Hua Ko which is a subsidiary of Hua Yung which is owned by the National Light Industries Corporation of China.

Hua Ko has a Californian subsidiary called Chipex and engineers from China working for Hua Ko are routinely posted to Chipex to study semiconductor technology in Silicon Valley.

The other company which is thought to be controlled from Peking - though it has never admitted that the financial backing is ultimately Chinese - is Elcap. Like Hua Ko, Elcap has invested in Silicon Valley and has a one fifth stake in Universal Semiconductor Inc of San Jose.

Hong Kong is regarded as a global 'staging post' for chips all over the world

None of the three chip makers of Hong Kong show up in the market analysts' lists of semiconductor producers so their size is unknown. As well as these chip manufacturing and assembly activities, Hong Kong is regarded as a global 'staging post' for chips from all over the world.

The fact that no major chip company has ever set up a full chip manufacturing facility including wafer processing in Hong Kong is generally attributed to the fact that the local government has demonstrated no coherent hi-tech policy which might encourage such firms to do so.

SINGAPORE

With Malaysia and the Philippines, Singapore is one of the world's three biggest assemblers of semiconductors. Unlike the other two, Singapore has specialised in testing as well as assembly which adds value to the end product.

The first companies to assemble in Singapore were Texas, Fairchild and National in 1968. Others like Hewlett-Packard and Teledyne followed. But local operations have not sprung up except for National Engineering Services.

Singapore is one of the world's three biggest assemblers of semiconductors

In 1979, the government of Singapore decided to encourage technology industries with a series of fiscal incentives for companies which upgraded the capabilities of their operations and the skills of their employees. This policy encouraged the semiconductor multinationals with assembly plants in Singapore to add testing facilities.

The government also encouraged the development of a skilled workforce by promoting vocational training institutions with the help of foreign governments and companies e.g. the Tata-Government Training Centre, the Philips-Government Training Centre, the Institute of Systems Science (IBM), the Japan-Singapore Training Centre, the Japan-Singapore Institute of Software Technology, the French-Singapore Institute of Electrotechnology and the Computervision-Government CAD/CAM training centre.

In 1981 the government tried to stimulate the pursuit of R&D in Singapore with a $23.7m grant towards semiconductor research projects. However, except for SGS-Thomson, no company had gone in for wafer processing in Singapore by the end of the Eighties.

THE PHILIPPINES

The second largest country for semiconductor assembly in the world, the Philippines entered the semiconductor industry in 1974 when Intel set up a plant there. Later on came Texas, AMI and Fairchild.

Sprague formed a joint venture with a local company called Deltron. Oki of Japan set up an assembly operation called Stanford Microsystems. Analog Devices and Honeywell both added test to their assembly operations.

The Philippines is the second largest country for semiconductor assembly in the world

However the mainstream semiconductor assembly industry of the Philippines has developed in a different way to other countries in that it has set up a large number of locally owned operations that assemble on a sub-contract basis.

The list of Philippines locally-owned semiconductor assembly companies is impressive: Asionics, Complex Electronics, Dynetics, Epic Semiconductor, Filipinas Microcircuits, Fruehauf Electronics, Integrated Circuits, Integrated Microelectronics, Nutech Circuits, Semiconductor Devices and Silicon Technology.

Dynetics also owns the Asian Reliability Company (ARCI) which in turn owns both the Manila-based Asia Test which provides testing facilities, and the Silicon Valley-based Tool and Die Testers.

The government has sought to encourage the industry by various means. In 1982 it introduced its 'no strike' law for semiconductor industry workers which makes arbitration compulsory in the event of labour disputes. In the same year the semiconductor industry was exempted from the surplus-profits tax and sub-contractor's tax which both apply in other industries. The semiconductor companies are also exempted from the law which applies elsewhere that companies must feed their workers.

The Philippines government gives a seven year tax holiday to semiconductor companies and this holiday can be extended if the company keeps upgrading the technological capability of its facilities. The government also bears the expense of training workers for the semiconductor industry.

MALAYSIA

The number one country for semiconductor assembly in the world with many US chip companies having assembly plants there: first in were National, Texas and AMD in 1972; the following year Intel, Motorola and Mostek went in; in 1974 RCA and Harris of the US, Hitachi, Toshiba and NEC of Japan and Siemens of Germany all set up assembly operations there.

Throughout the Eighties semiconductor assembly represented the largest manufactured export from the country. Although test and other back-end processing functions are increasingly being undertaken in Malaysia, the government does not appear to have planned for the country to be anything other than a low cost host for foreign companies' subsidiaries.

OTHER COUNTRIES IN THE SEMICONDUCTOR INDUSTRY

Other countries also in the semiconductor industry with assembly operations are:

Mexico has a local sub-contract operation called IMEC and the assembly subsidiaries of Motorola, Fairchild, Intel, International Rectifier and Solitron Devices of the US and Toshiba of Japan.

Brazil has Ford Philco of the US, Siemens of Germany, Philips of Holland, NEC and Sanyo of Japan. The government has tried to encourage local activities with initiatives which have not always been well-judged.

Thailand has two local companies Sea Electronics and Sidha International. Foreign companies there include National, Data General, and Honeywell.

So that's the semiconductor industry for you - a sprawling global industry molly-coddled and courted by governments as the key provider of enabling technology for the electronics industry. To many countries prowess in semiconductors is a measure of its sophistication as an industrial country.

The semiconductor industry is a sprawling global industry molly-coddled and courted by governments as the key provider of enabling technology for the electronics industry

CHAPTER 7

WHO ARE THE PLAYERS?

The biggest worldwide companies; the biggest company in each of the '4Cs': computers; communications; consumers; components. Who they are; how they got there and what they do.

THE BIGGEST WORLDWIDE COMPANIES

The 20 largest manufacturers of electronics goods in the world are: IBM (US), Matsushita (Japan), NEC (Japan), Philips (Holland), Toshiba (Japan), Hitachi (Japan), Fujitsu (Japan), CGE (France), Siemens (Germany), General Electric of the US, Alcatel NV (Holland), DEC (US), General Motors (US), Xerox (US), AT&T (US), Unisys (US), Hewlett-Packard (US), Motorola (US), GEC of the UK and Thomson of France. Annual sales range from $45bn to $6bn. That's nine US, six European and five Japanese companies. (See also page 141.)

General Motors may seem an odd name there but it got into electronics in the Eighties by buying Hughes' electronics interests as well as a load of companies involved in producing automated factories. It is these which give it a late Eighties turnover of around $11bn in electronics products.

These are not the biggest corporations in the world involved in electronics. If you looked at it on that basis General Motors, a $100m+ company, would head the list. The ranking here is purely on the basis of sales of exclusively electronics products.

To get the top 20 American companies you add to the nine in the world top 20: NCR, Honeywell, Texas Instruments, Raytheon, Lockheed, Apple, Tandy, Westinghouse, Control Data, Rockwell, Wang, Litton Industries and TRW.

To get the top ten Japanese companies you add NTT, Oki, Sanyo, Sony and Mitsubishi to the five Japanese companies listed in the world top 20.

And to get the top ten European companies you add STC of the UK, Olivetti of Italy, Ericsson of Sweden and Bull of France to the six other European companies in the world top 20.

One thing stands out - the Americans have more large companies than anyone else. At the end of the Eighties the US boasted over 50 companies with electronics sales worth over $1bn a year - more than three times the number in Europe and more than twice the number in Japan.

The Americans have more large companies than anyone else - more than three times the number in Europe and more than twice the number in Japan

The Interesting Origins of Some of the Top Companies

Many of the top companies have interesting origins:

- IBM developed from the Computing Tabulating Recording Company of the punched card pioneer Herman   Hollerith;

- AT&T derived from the commercial operations of Alexander Graham Bell;

- GE of the US was the company to which Edison sold out the commercial interests surrounding his 'invention factory';

- NEC of Japan started as a joint venture with AT&T and spent its first 33 years under US management control;

- Philips of Holland provided much of the initial technology which allowed Konosuke Matsushita to found the company which became, in his lifetime, the largest electronics company in Japan and the second largest in the world. In tribute, a statue of Philipsco-founder Anton Philips stands outside the Matsushita HQ in Osaka;

- at the end of the Eighties, DEC is still run by one of the men who founded it in 1957 - Ken Olsen;

- Sony is still headed by Akio Morita who founded the company in 1945;

- Two other leading entrepreneurs were Bill Hewlett and David Packard who grew the company they founded in 1939 - Hewlett-Packard - to be the eighth largest electronics company in America by the end of the Eighties.

In components, the notable entrepreneurial successes are Intel, National Semiconductor and AMD. Twenty years on from the late Sixties when these companies were started up, their founders still run them: Robert Noyce, Gordon Moore and Andy Grove were still, at the end of the Eighties, heading Intel which reached $3bn annual revenue mark at the end of the Eighties; Jerry Sanders still headed up Advanced Micro Devices, and Charlie Sporck was still running National Semiconductor.

Intel, National, Texas, Motorola and AMD are all in the US top 50 and are the only components manufacturing companies, as opposed to equipment manufacturing companies, to figure in the list.

To put some shape round this discussion of the players it makes sense to look at the top notchers in each of the 'Cs' - Computers, Communications, Consumer Electronics and Components.

Broadly speaking you can say that the consumer business is owned by Japan with a threat from Korea; the computer business is owned by America with a threat from Japan; the communications business is split between the US and Europe, and the components business is split between America and Japan.

COMPUTERS

The World Leaders

In computers, at the end of the decade, the leaders are, in descending order of size: IBM, DEC, Fujitsu, NEC, Unisys, Hitachi, Hewlett-Packard, Siemens, Olivetti (Italy) and NCR (US).

IBM's revenues at $55bn are equal to the combined revenues of the next six largest players. It has been said that IBM is under threat as the big computer gives way to the desktop computer which almost anyone can make with parts bought off the shelf. But IBM is still the main supplier of desktops even if it is losing market share.

> **IBM's revenues at $55bn are equal to the combined revenues of the next six largest players**

For many years IBM was the world's largest manufacturer of chips however in 1989 NEC of Japan is said to have matched the IBM output which is estimated to be worth $3.6bn. IBM uses all its own chips and is thought to have more advanced R&D in chip technology than anyone else. As well as its edge in silicon technology, IBM has an edge in that it sets the standards to which the industry conforms. While it retains both those positions it will probably hold on to its pre-eminence. IBM is market leader in every capitalist country in which it operates. However, the Japanese players, Fujitsu, NEC and Hitachi are steadily making ground.

DEC, the number two player is big in the minicomputer business which it invented; Unisys - the merged Burroughs/ Sperry - is a broad range supplier like IBM; Hewlett-Packard is making ground on the back of a new concept in microprocessor designs (called RISC - Reduced Instruction Set

Computing) which was introduced in the mid-Eighties; and NCR is big in banking and retailing systems as its full name 'National Cash Registers' implies.

Fujitsu both sells its own-brand machines and supplies the electronic guts of machines sold by Siemens of Germany and ICL of the UK. In the US, Fujitsu sells machines through the US company Amdahl in which it has a major stake, and Hitachi sells its machines through a company called Comparex under the latter's name.

The two European players in the top ten, Olivetti and Siemens, differ fundamentally in that the Italian company makes desktops, while the German company makes a broad range of machines including mainframes. Both make machines which conform to the standards set down by IBM.

The Future of the Industry

So one of the interesting questions about the industry in the Nineties will be 'Can IBM continue its reign?' The Japanese are after it; the personal computer companies are after it; the Europeans are trying to match its silicon capabilities; challenges to its standards are being thrown down.

'Can IBM continue its reign?' - the Japanese, the personal computer companies and Europeans are challenging it

In the electronics industry IBM remains the most potent symbol of US supremacy. It has saved US technology in the past and is saving it again - coming to the rescue of the beleaguered US chip industry in 1989 with the offer of its 4 Megabit DRAM and saying it will give the technology to US manufacturers if they combine to set up a factory to make it.

But even IBM is subject to the market and the corrosive effect of falling profits. If it ever finds that its diminishing share of the PC markets is pushing it towards loss, the expensive programmes by which it retains its technological eminence will be under review.

COMMUNICATIONS

The Key Companies in a Period of Fast Change

Manufacturing communications equipment is a smaller business than manufacturing computers. The top ten companies' communications revenues range from $10bn to $2bn. The list runs AT&T, Alcatel NV (which is a joint venture between the CGE subsidiary Alcatel and ITT of the US), Siemens, NEC, Northern Telecom (of Canada), IBM, Motorola, Ericsson (of Sweden), Fujitsu and Philips.

The real money in telecommunications - said to be worth some $250bn a year worldwide - comes from operating the networks which take money off people making telephone calls.

> The real money in telecommunications comes from operating the networks which take money off people making telephone calls

The changes in telecommunications have all come about as a result of 'deregulation' i.e. getting rid of state control of telephone networks and state control of who supplies equipment to be used on telephone networks.

This has meant that the old 'Cosy Club' of national suppliers - AT&T in the US (both a supplier and a network operator), Italtel in Italy, GEC and Plessey in the UK, Siemens in

Germany, Alcatel in France have all had to face up to foreign competition.

This has resulted in alliances - Siemens of Germany and GTE of the US, Alcatel of France and ITT of the US, Fiat's Telettra and Italtel of Italy, GEC and Plessey. It has also resulted in a host of new styles of telephones, and other kit which can be attached to the telephone plug - answerphones, facsimile machines, cordless phones etc.

Now you no longer have to rent standard model telephones from a monopoly state-approved supplier. Instead, the race is on for who can produce the cheapest and/or most innovative products which the public want just as in any consumer electronics market.

The biggest revolution of the second half of the Eighties is, however, the cordless, mobile telephone which precedes a range of cordless equipment like facsimile. This is going to be a big future market. Already people see pocket phones becoming as cheap and as plentiful as calculators. Telecommunications is on the move from the staid old auntie public utility offering jobs for the boys, to a whizzy hi-tech consumer industry offering the benefits of cut-throat competition to the consumer.

Telecommunications is on the move from the staid old auntie public utility offering jobs for the boys, to a whizzy hi-tech consumer industry offering the benefits of cut-throat competition to the consumer

CONSUMERS

The Leading Companies and the Dominance by Japan

The top companies in consumer electronics are, in descending order of size: Matsushita and Sony of Japan, Philips of Holland, Toshiba and Hitachi of Japan, Thomson of France, Sanyo, JVC and Sharp of Japan, Bosch and Grundig of Germany and Nokia of Finland.

Japan also has two top specialist consumer companies - the watch kings Citizen and the calculator kings Casio. The US has only two remaining significant consumer players left in ITT and Zenith.

Although the Japanese have built up their electronics industry since the war very largely from the profits earned by selling electronics consumer goods abroad, there is evidence that their market share in Europe has peaked.

In significant areas like TVs and VCRs the Japanese share of the market has remained unchanged during the second half of the Eighties. To an extent that is a result of the rise of Korea in consumer electronics but it is also due to the renewed European capability.

We are seeing European companies being driven out of their traditional reliance on military and telecommunications markets by defence cut-backs and deregulated telecoms markets and struggling to get back into consumer electronics.

European companies are being driven out of their traditional reliance on military and telecommunications markets and are struggling to get into consumer electronics

An example is the resurgence of consumer activity at Thomson of France, which had a notable innovation in the magnetic-optical disc, and which in 1989 bought both the US consumer group RCA and the UK consumer group Ferguson.

Another example is at Nokia of Finland which aims to build itself into a major world player in consumer electronics. It has surged to become the ninth largest world TV producer by the end of the Eighties.

Europe's outstanding player in consumer is, however, Philips of Holland. Inventor of both laser and compact disc technology, Philips is the prime innovator and European presence in the consumer industry.

The opportunity to get back into consumer electronics is seen, both in Europe and by the US, as high definition TV (HDTV). What HDTV is sounds simple. If you look closely at a TV picture you see it is made up of horizontal lines and dots. Obviously, if you double the number of lines and halve the size of the dots you'll get a better picture. That's what HDTV aims to do.

The US, Japan and Europe are pouring billions of dollars into it to get the technology to the point where it is affordable to the consumer. The EEC funded ESPRIT R&D programme is making high definition TV one of its highest spending projects.

The Nineties looks like being a decade where the US and Europe slug it out with Asia to win in the consumer electronics markets

Backing these attempts in the equipment area are programmes in the chip area worth several billion dollars to provide the underlying technology. So the Nineties looks like being a decade where the US and Europe slug it out with Asia to win in the world consumer electronics markets - to the benefit of all of us.

COMPONENTS

Lastly components, but it is by no means least! The manufacturing of components accounts for between a quarter and a fifth of the total value of the electronics industry. For passive components the industry is mature, but for actives it is far from mature with the chip industry still maintaining its average year by year growth of 20%.

At the end of the Eighties there were only about 100 companies in the world making chips which had revenues of $5m or more:

A Breakdown of Output

- 60 American producing about 40% of the industry output;

- 16 Japanese producing about 50% of the output;

- 12 European producing about 8% of the output;

- and the rest produced around 2% of the output.

The Main Companies

In the US the Big Three - all in the world top ten - are Texas Instruments, Motorola and Intel. Texas and Motorola started making semiconductors in the Fifties, Intel was one of the Sixties generation. Next in the US top league are the $1bn companies National Semiconductor and Advanced Micro Devices.

They have survived because they have retained a lock on important product areas. Texas invented and still dominates the market for commodity logic chips. Intel and Motorola still dominate the market for microprocessors. National dominates the market for linear chips. AMD's strength is in bipolar logic, particularly in a super-fast bipolar microprocessor called a 'bit-slice', which it invented.

The Big Three - all $3bn companies - look secure. But others who started up in the Sixties like National and AMD are having difficulty maintaining growth and profitability because the areas which they dominate - linear and bipolar logic - are slow-growth areas of the market and are not keeping up with the overall market growth.

Although these companies have moved into new areas - like the sudden Eighties switch into CMOS technology - they have found these areas ferociously competitive with a mass of hungry new start-up companies founded in the Eighties specifically to take advantage of the technology shift.

The most successful of the new companies are: the 1979 start-ups Micron Technology and LSI Logic which were $400m companies at the end of the Eighties; the 1978 founded Integrated Device Technology, now a $200m+ company; the 1985 start-ups include Chips and Technologies, also now a $200m company - the fastest growing in the industry's history; and the 1983 start-up Cypress Semiconductor which is a $150m+ company.

Just as Fairchild, Motorola and Texas pushed out the old Fifties leaders like Transitron, Hughes etc., and Intel, National and AMD pushed out the likes of RCA and General Electric so the new young companies claim they will take the place of National and AMD.

Meanwhile National and AMD have looked for salvation via mergers - National with Fairchild, AMD with Monolithic Memories. But the companies founded in the Eighties would say that decline and renewal are the natural cycle of the US semiconductor industry and that their time has come.

The Components Industry Worldwide

Only in America will true free market economics produce this kind of lively, changing domestic industry where the fresh and the fit have the chance to oust the old and ineffective.

But even in America, the bastion of laissez-faire economics, we are now seeing the US Congress funding a catch-up chip technology consortium. Called Sematech, it is made up (with one exception) of the 20 year-old and older companies. Sematech is seen by the newer, livelier companies as a prop-up programme for ageing, ailing companies which have run their course.

America's chip industry remains the world's most interesting, mobile and innovative. It is the only place on earth where a chip engineer with a new idea can get venture capital funding and turn himself into a millionaire. New start-up chip companies are still appearing there as plentifully as ever. By comparison the structure of the rest of the world's chip industry is static.

America's chip industry remains the world's most interesting, mobile and innovative - by comparison the structure of the rest of the world's chip industry is static

In Japan the chip business is dominated by the semiconductor divisions of the large equipment companies - NEC, Toshiba, Hitachi, Fujitsu, Mitsubishi and Matsushita all of whom are in the world top ten for semiconductors with a second team of Sanyo, Sharp, Sony, Rohm, Sanken, Seiko, Fuji Electric, NMB, Yamaha and Ricoh.

Although the history of the industry shows collaborative efforts to develop technology, companies have retained their technological specialities: Toshiba has the world's best CMOS process; Hitachi has the world's leading BiCMOS process; Fujitsu and Sony have exceptional bipolar processes.

The history of the electronics industry shows collaborative efforts to develop technology, but companies have retained their technological specialities

The chip divisions of the Japanese Big Three are all around the $4bn mark, and the chip operations of the next three biggest are in the $2bn to $3bn range. By concentrating on the largest segment of the chip market - memory chips - they have achieved a larger market share than the Americans and threaten to knock them out of the industry.

That's because only the revenues obtainable from selling memory chips are big enough to cover the cost of the basic R&D needed to stay in business. Hence the late Eighties joint ventures of Hitachi/Texas and Toshiba/Motorola both concentrated on memory technology exchange.

The Japanese are now preparing to go for the throats of the US Big Three. They are gearing up in microprocessor manufacture which will hit Intel and Motorola and they are seeking to wipe out Texas' dominance in bipolar logic by replacing it with CMOS logic in which they have an edge.

Japan makes over half the world's microchips, and dominate the chip industry - if they can get control of the computer industry, they will dominate the world industrially in a way no country has ever done

So the Japanese, already making over half the world's output of microchips, look like dominating the world chip industry. If they achieve a lock on that market they will then be able to go on and wipe out the US position in computers. If they do that, they will dominate the world industrially in a way no country has ever done before.

However there are extra-industry factors and pressures which may prevent that happening. External political pressures are pushing Japan to locate factories abroad, to do R&D abroad, to form alliances with foreign companies, to exchange technology and to generally trade-off some of its competitive edge for goodwill from its trading partners. How it all turns out will be one of the interesting scenarios of the Nineties.

In Europe the semiconductor industry is also dominated by large companies - Philips, SGS-Thomson and Siemens, with a second team of Plessey (jointly owned since 1989 by GEC and Siemens), Matra-Harris, Telefunken, Ericsson, Marconi, Mietec, STC and European Silicon Structures.

The European Commission seems to have now adopted it as a policy that the European semiconductor industry will be nurtured with subsidies, protected by tariffs and fed with preferential local procurement status.

When, at the beginning of the Eighties, it looked as if the European car industry would be wiped out by the Japanese, European governments decided to stop that happening. They succeeded. They can do the same with semiconductors if they so wish.

In non-Japanese Asia, the semiconductor industry is dominated by the Korean company Samsung - the only non-Japanese billion dollar chip company in Asia at the end of the Eighties. Samsung has shown it will spend whatever it takes to become a top ten player in the semiconductor industry.

An Asia/Pacific second eleven consists of: United Microelectronics Corp, Taiwan Semiconductor Manufacturing Company, Hualon Microlectronics Corp, Winbond Electronics, Advanced Microelectronic Products and Vitelic of Taiwan; Lucky Goldstar, Hyundai, KEC (Korean Electronics Company) and Daewoo of Korea; RCL, Elcap, Hua Ko and Micro Electronics of Hong Kong,

The strongest growing of these are UMC and TSMC of Taiwan and Hyundai and Lucky Goldstar of Korea. The others will stay second rank for the foreseeable future. The Hong Kong companies will continue to act more as a technology conduit for China than companies seeking revenues, profits, market share or size for their own sake.

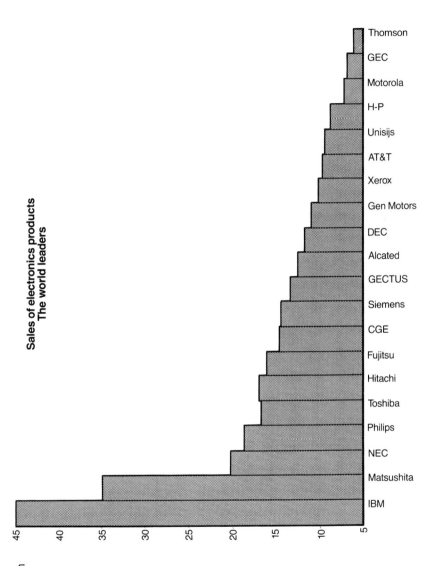

Sales of electronics products
The world leaders

$ billion

CHAPTER 8

WHERE DOES IT ALL HAPPEN?

The 20 biggest electronics producing countries on earth from America to Malaysia.

Because the electronics industry has many multinational companies - like IBM, DEC, Philips, Hitachi or NEC -which produce and sell in many different countries, the distinction between what is 'Korean' or what is 'Canadian' when speaking of the manufactured output of any particular country becomes blurred.

> **The electronics industry has many multinational companies, which produce and sell in many different countries, so the manufactured output of any particular country becomes blurred**

In each country the multinationals mix with locally owned firms to add value in that country. It is on that basis - not on the output of 'British firms' or 'Brazilian firms' that the following list of the largest electronics producing countries is drawn up.

At the start of the Nineties, the world's 20 largest national producers of electronics goods are, in descending order: America, Japan, Germany, France, Britain, Italy, Korea, Taiwan, Singapore, Holland, Canada, Brazil, Switzerland, Sweden, Hong Kong, Belgium, Spain, Ireland, India and Malaysia.

AMERICA

With over 20,000 electronics companies, more than 50 of which have over $1bn in annual sales, and with massive government budgets for defence and space electronics, the USA is the strongest electronics nation on earth.

We saw the list of the 20 biggest companies in America in Chapter seven. The list of the next 30 reads: Litton Industries, TRW, ITT (though a $20bn company ITT gains only $2.7bn from its electronics products), National Semiconductor, Zenith, Intel, AMP, Harris, Honeywell Bull, Schlumberger, du

Pont de Nemours, GTE, Compaq, Amdahl, Avnet, Prime, Teledyne, Eaton, Tektronix, Data General, Perkin-Elmer, Seagate, Tandem, Varian, AMD, E-Systems, United Telecom, Northrop, Sun and Ford Aerospace. Size ranges from $3bn to $1bn.

The main areas are computers, communications and components. Consumer electronics was a US capability which collapsed under attack from Asia leaving US electronics consumer imports worth over ten times more than the value of its consumer exports.

> **The main areas of US production are computers, communications and components - their consumer capability collapsed under attack from Asia**

Currently under pressure is the components sector. We have already seen the effect which that pressure is having on the US semiconductor companies - forcing them to join in consortia supported by the taxpayer which is a very un-American activity indeed.

So whether or not the Japanese can send the US components industry down the same plughole as the consumer industry will depend very much on the attitude of the US government.

About half the US communications market is based on sales to the US government, of which 40% is military. The top ten suppliers to that market are: Honeywell, Hughes, Martin Marietta, Motorola, Raytheon, RCA, Rockwell, Sperry, Texas and United Technologies.

Deregulation of the civil communications market has let in foreign suppliers and the most notable exploiters of this situation are Northern Telecom of Canada, Plessey (via its purchase of Stromberg-Carlson) now part of GEC/Siemens, and Ericsson of Sweden. So far the Japanese have not made a major play.

In the computer area, the Japanese are beginning to make big inroads in PCs but not yet in big machines though, as we saw in Chapter seven, Fujitsu and Hitachi supply these through the US companies Amdahl and Comparex.

JAPAN

In 1990 Japan became the first country in the world in which electronics was the largest industry - surpassing cars, textiles, steel, chemicals etc. The country, world number two in electronics output, exports half its electronics production which represents about 20% of Japan's total exports.

> In 1990 Japan became the first country in the world in which electronics was the largest industry

The major companies are household names and most have been mentioned in previous chapters - Matsushita, NEC, Toshiba, Hitachi, Fujitsu, Mitsubishi, Canon, Oki, Pioneer, Ricoh, Sanyo, Sharp, Sony, Seiko Epson, Casio, Citizen, Fuji Electric, JVC, Orient, Rohm, Sanken Electric, Aiwa, Akai and Yamaha.

Unlike in the US, where companies tend to specialise in one of the four 'Cs', in Japan the same companies tend to dominate right across the industry. For instance, of the six companies in the world top ten for semiconductors (NEC, Toshiba, Hitachi, Fujitsu, Mitsubishi and Matsushita), three show up in the world top ten for consumer electronics (Matsushita, Toshiba and Hitachi), and four show up in the Japanese top ten for computers (NEC, Hitachi, Toshiba and Mitsubishi).

Which gives the clue to the nature of the Japanese electronics industry - big, vertically integrated companies dominate it. Adding even greater strength to these already huge companies is their practice of clustering in supportive groups of other companies around a large bank. This gives them great powers of survival through bad times.

The principal sectors of the Japanese electronics market are components (making up about 30% of the total Japanese electronics production), consumer (25%) and computers (20%).

Forty years of unending growth and expansion in world markets have led the big Japanese companies to expect success and there is not much that can dent their confidence.

Inroads into the Japanese position may come through two political moves: first the mid-Eighties privatisation of the monopoly telecommunications operator NTT was supposed to herald the opening up of Japan's telecommunications equipment market to foreign suppliers; and second, the 'Chip Wars' of the Eighties fought against the US and EEC were supposed to have been settled on the condition that Japan opens up its components market. The Nineties will see the proof in these particular puddings.

The only things that seriously worry the Japanese are Western trade restrictions, the emerging Koreans and later the possible emergence of the Chinese.

Already they are increasingly putting the low value-added end of their operations off-shore and are being forced to locate high-added value elements of their operations in the markets which they serve. Trade issues and Asian competition appear to be the only factors which the Japanese seem to regard as hindrances to their global domination of the electronics industry.

Trade issues and Asian competition appear to be the only factors which the Japanese seem to regard as hindrances to their global domination of the electronics industry

WEST GERMANY

Germany is Europe's largest electronics producing country, and the world's third largest. It has a diversified electronics industry which is strong in the main areas of computers, components, communications and consumer, and particularly strong in the control and instrumentation area.

> Germany is Europe's largest electronics producing country, and the world's third largest

The $30bn giant Siemens dominates the German electronics industry. The other major companies are: AEG-Telefunken, Nixdorf, Bosch, Rohde and Schwarz, Grundig, Diehl, Bergmann, BASF, Mannesmann and Hoechst.

At the end of the Eighties Siemens is making a big push to catch up with the Japanese in chip technology. The early Nineties will be the judge of whether the $2bn or so spent on this project will have been wasted. A joint venture with AEG to produce next generation chip-making equipment was said, in 1989, to have a two year lead on the Japanese.

FRANCE

France comes fourth in the world production league though some way behind Germany. France is the strongest telecommunications company in Europe. Telecoms and military account for 40% of the country's electronics output. It is beginning to find its military sales to the Third World are reducing which is causing the defence industry to retract.

The top French companies are: CGE, Bull, Matra, Thomson, Souriau, Croujet, Jeumont-Schneider, Telemecanique Electrique, TRT, SAGEM, SAT and CSEE.

As we saw in the Chapters five and six, the French government is very interventionist and more effectively so than in Britain promoting major restructuring exercises like:

The Government's Restructuring Exercises

- CGE merged its telecommunications subsidiary, Alcatel, with ITT's worldwide telecoms business in 1986;

- Thomson absorbed the consumer electronics activities of RCA which had been previously bought by General Electric (of the US);

- Thomson also bought Ferguson the consumer electronics division of Thorn EMI;

- Thomson also merged its semiconductor division with SGS of Italy;

- the French telecoms company CGCT was absorbed into the Swedish Ericsson which in turn linked with the French Matra.

These deals were in some respects the brain-children of government. The government has also encouraged the inward investment into France of over 100 major foreign companies.

The French industry breaks down: 40% telecoms/military, about 23% computer and 20% components.

BRITAIN

Britain is fifth in the world electronics production league. Its biggest companies in the business - GEC, STC, Plessey, Racal and Thorn had electronics sales worth some $7bn, $4bn, $2.5bn, $2bn and $1.5bn respectively in 1989. Rationalisation in 1989 saw GEC and Siemens of Germany take over Plessey and divide it up with the larger share going to Siemens.

Apart from these, the UK's larger electronics concerns include: Ferranti, British Aerospace, Dowty, Amstrad, Bowthorpe, Oxford Instruments, BSR, SD-Scicon, Sema Group, Lex Service, Logica, Gestetner, Smiths Industries, Rank, Cambridge Electronic Industries, BET, Cambridge Instruments, Crystalate, Apricot.

Overseas companies contribute heavily to UK output. IBM (UK) has an output worth some $3bn a year and Philips' UK activities have sales worth around $1bn a year.

Three-quarters of the output of the UK electronics industry is for the professional electronics equipment sector including defence equipment, telecommunications equipment and computers. The rest of the UK industry is made up 17% components and about 6% consumer.

The British Government gives a good support for military R&D but otherwise its efforts with the electronics industry have been ineffective

The government gives a good deal of support for military R&D but otherwise its efforts to co-ordinate industry strategies and technological development have been relatively ineffective.

Telecommunications has been liberalised with a second network operator called Mercury being licensed to challenge the former monopoly operator British Telecom. Also many licences are being given out to mobile phone companies and a policy of open procurement for mobile phone equipment has led to Britain having the cheapest and most competitive mobile telephone market in the world at the end of the Eighties.

Britain is also at the forefront of introducing new forms of mobile communications such as the one-way (make but not receive) digital system called 'Telepoint' using so-called 'CT2' telephones (stands for 'Second Generation Cordless Telephone'). Britain is also introducing a two-way digital system called 'Personal Communication Network'.

These networks should open the way to highly competitive equipment markets which will force down the price of mobile telephones and calls.

ITALY

Italy lags some way behind Germany, France and the UK coming in sixth in the world league but with higher growth than many other European countries (5.6% against the European average of 4%) except for Spain (on 6.2%).

The main companies involved are IRI-STET (a State-owned holding company which owns the major telecommunications company Italtel and half of the Italian/French semiconductor company SGS-Thomson); Olivetti, Zanussi, REL and Fiat.

Italy's government has also been instrumental in promoting rationalisation of the Italian electronics industry. In the late Eighties, Fiat's telecommunications subsidiary, Telettra, merged with Italtel forming a single company called Telit.

SGS-Thomson - the merged activities of the semiconductor subsidiaries of Italtel and Thomson and Europe's second largest semiconductor company - will receive substantial funding during the Nineties under the various EEC microelectronics initiatives.

The Italian government has a ten year telecommunications plan (1982-92) to improve technical capability and invests substantially in chip R&D to SGS and to universities.

Computers and industrial automation are the two other major segments in the Italian electronics industry.

KOREA

Next largest electronics producing country, and the largest outside the US, Japan and Europe, is Korea. Throughout the Eighties, Korea's electronics industry grew at around 12% per year.

The Korean government directs and co-ordinates the electronics industry's efforts to focus on particular industry segments. The initial strength of Korea was in consumer electronics but throughout the Eighties it focused on semiconductors and rapidly increased both technical capability and exports. The next areas to focus on will be telecommunications and computers.

The Korean Government directs and co-ordinates the electronics industry's efforts to focus on particular industry segments

The Korean telecommunications network is being rapidly expanded and digitalized and co-operative efforts by companies and universities to pursue computer R&D are being encouraged by government subsidies.

The electronics industry is dominated by four conglomerates or 'chaebols' as they are called: Samsung, Lucky Goldstar, Hyundai and Daewoo. Other substantial companies are Anam, the largest semiconductor assembly house in the world, Dongwon Electronics, Korean Semiconductor, Sunglee, and Tae Kwang Industrial.

TAIWAN

Eighth in the electronics world league, but growing rapidly, comes Taiwan. In keeping with the Chinese traditions of entrepreneurship, there are many small companies. More than half Taiwan's 2,700 electronics companies have sales under $1m.

70% of the country's production is exported of which over half goes to the USA. Consumer electronics is Taiwan's greatest strength. Monitors and personal computers are the biggest single products exported.

> 70% of Taiwan's production is exported, of which over half goes to the USA

Major companies are Asa, Tatung, Calcomp Electronics, Kuo Ferg, United Microelectronics Corporation, Shirlin Electric and Engineering and AOC International.

SINGAPORE

Singapore's electronics output is nudging the size of Taiwan's but with a very different structure - 250 companies compared to Taiwan's 2,700. Singapore's largest industry sector is components followed by consumer and industrial.

Major companies are Singapore Semiconductor, Advanced Concept Technology, ASJ, Neutron, Setron and Mayentro Industrial.

The government has promoted Singapore actively as a suitable location for foreign companies. Hewlett-Packard, NEC, Texas, Siemens, SGS, Philips, Sanyo, National Semiconductor, Fujitsu, Compaq, Apple, GE of the US, AT&T, Nixdorf and Aiwa have all responded.

HOLLAND

Tenth in the world electronics production league is Holland. The country's electronics industry is dominated by Philips which is Europe's only genuine multinational electronics company.

> **Holland's electronics industry is dominated by Philips - Europe's only genuine multinational electronics company**

If the $26bn Dutch colossus was a country doing all its manufacturing within its own boundaries, it would be in the top half dozen electronics producing countries.

In fact it manufactures in 48 and markets in 90 countries. It is the sixth largest electronics company in the world, according to Fortune and the twenty-second largest industrial company overall. The company made its name in lighting equipment which still accounts for over a tenth of its sales.

Unlike many of the West's electrical/electronic companies it has never abandoned consumer electronics which account for about a third of its sales. Professional electronics is responsible for another third and components for about an eighth of overall revenues. The company is involved in every sector of the industry, as it is also in telecommunications, defence electronics and personal computers.

The telecommunications market in Holland is supplied by Philips and Ericsson but this market is to be opened up to others during the Nineties following privatisation of the network operator.

CANADA

Canada comes in eleventh though the country has a large trade deficit in electronics goods. Canada's electronics industry is dominated by foreign companies.

The Canadian electronics industry is unbalanced with telecommunications being by far the largest sector accounting for a third of all electronics production. Northern Telecom is Canada's only world-class electronics company. Mitel was another but it is now controlled by British Telecom.

After telecoms comes the computing industry led by Northern Telecom again, AES Data and Gandalf. Computing accounts for about a fifth of electronics output. Components - PCBs, connectors, power supplies - account for about 10% of output.

BRAZIL

The Brazilian electronics industry has been heavily directed by government even to the extent that its computer industry development policy is laid down by law.

Brazil's computer industry development policy is laid down by law

The manufacturing of computers and peripherals has been growing at 30% a year from the mid-Eighties onwards. Largest of these are Cotra Computadores and Labo Electronic.

The major telecommunications companies are Daruma Telecommunications and Elebra.

SWITZERLAND

Thirteenth in the world league, is Switzerland. Major Swiss companies involved are Brown Boveri (now merged into the Swedish group Asea), Hasler and Autophon, (now merged into the group Ascom) Ciba-Geigy, Landis & Gyr, Oerlikon Buhrle, Zellweger Uster and SMH.

Four of these, Brown Boveri, Zellweger, Ascom and Landis and Gyr have combined to form Dectroswiss - a joint initiative in chips. SMH has for many years run a chip plant at Neuchatel and licenses technology to run it from Intel and VLSI Technology of the US.

The strongest part of the Swiss electronics industry is control and instrumentation products particularly for air conditioning, ventilation and heating. Deregulation of the telecommunications industry is promised for the Nineties.

SWEDEN

Sweden's electronics industry is characterised by the domination of a few very large companies like Asea, Ericsson and Electrolux. Large companies account for some 90% of Sweden's electronics output by value.

The government is spending heavily to promote R&D and has promised deregulation of the telecommunications industry and a new network operator called Scantel to compete with the old monopoly operator Televerket.

HONG KONG

Consumer is the name of the Hong Kong game - watches, clocks, radios, calculators. Main local companies are Conic Elcap, Elec and Eltek, Nam tai and RCL Semiconductors.

Hong Kong's electronics industry is characterised as more of a sub-contract industry than a place for the subsidiaries of major companies. This could be a result of the government's policy of general laissez faire to all industries which means that it does not offer the special grants and inducements which other governments generally offer specifically to electronics companies.

> Hong Kong's electronics industry is characterised as more of a sub-contract industry than a place for the subsidiaries of major companies

It is almost impossible to assess the value-added of the Hong Kong electronics industry as so much is coming in and out on a sub-contract basis and on an import-for-retail basis. It is put at around $4.5bn.

BELGIUM

Belgium has a number of subsidiaries of major foreign companies but has no major world-class electronics company of its own. However the Belgian engineering group ACEC has been trying to turn itself into more of a high-tech oriented company.

The bulk of the foreign-owned manufacturing in Belgium is of computers and telecommunications equipment. It is also the location of the chip-making subsidiary of Alcatel - Mietek.

SPAIN

Seventeenth in the world league, Spain has been trying to expand its electronics industry throughout the second half of the Eighties. Government plans focus on attracting foreign companies to produce goods in Spain which will cut down the country's dependence on electronics imports and create exports.

AT&T, H-P, Sony, Grundig, Sanyo, IBM, Olivetti, Fujitsu, Sperry and Sharp have all responded to the call.

More competition in the telecommunications sector is expected as La Telefonica - Spain's monopoly network operator owned 45% by the State, eases up on the conditions under which companies can supply equipment. National companies include Piher the components company and Amper the telephone equipment manufacturer.

INDIA

India's electronics production has been growing strongly throughout the late Eighties - tripling output between 1983 and 1989. The government targeted a 39% annual growth rate for the industry in the Eighties and backed it up with policies promoting the import of technology and hardware, funding R&D and allowing freedom to operate for companies with foreign shareholdings.

India's electronic production has been growing strongly throughout the late Eighties - tripling output between 1983 and 1989

The government has set up research centres for microelectronics, for materials used in electronics and for telematics. The government-run Semiconductor Complex is involved in R&D on leading technology chips helped by US companies like Rockwell.

The main industry areas in India are computers and telecommunications but in the late Eighties the government focused on promoting consumer electronics - watches, microwave ovens, VCRs, TVs etc.

The main consumer companies are Bajaj Electricals, Gramophone Company of India and Peico Electronics and Electricals. Leading components companies are Bharat Electronics, Central Electronics, Electronics Corporation of India and Hind Rectifiers. Other majors are Hindustan Teleprinters and Indian Telephone Industries.

SOUTHERN IRELAND

Nineteenth in the world electronics production league is Ireland. Nearly all of this comes from the subsidiaries of foreign multinationals attracted in by grants from the Irish government.

Many multinationals have subsidiaries in Ireland attracted by grants from the government

Notable is the computer sector with Apple, Wang, Amdahl, DEC, Nixdorf and Verbatim which accounts for over half of the country's electronics output.

Telecoms is well represented with Ericsson, AT&T and Northern Telecom. In components there are Fujitsu, NEC, GE of the US, Analog Devices, and Bourns. Electronics accounted for a quarter of Ireland's total exports at the end of the Eighties.

MALAYSIA

As we saw in Chapter six, Malaysia is the world's largest assembler of semiconductors which accounts for nearly 80% of the Malaysian electronics industry's total output of some $3bn which makes it the twentieth largest nation for electronics output in the world. The assembly industry consists of the subsidiaries of foreign companies.

THE REST OF THE WORLD

After the top 20 countries in electronics production, the involvement of the other countries involved declines in value of output quite quickly. The next ten in the world electronics league are: Austria, The Philippines, Finland, Israel, Denmark, Australia, Norway, Indonesia, Thailand and South Africa. With South Africa well below the $1bn total output mark at the end of the Eighties.

CHAPTER 9

WHERE ARE WE ALL GOING?

The trends for the Nineties: open systems;
compatibility; convergence of
telecommunications and computing;
globalisation; trans-border takeovers and
mergers; the decline of defence electronics;
systems on a chip; multiprocessing; networking
and the decline of the mainframe; the rise of the
cordless telephone; the pre-eminence of
software; transportability; technological
democratisation.

Identifiable trends in the electronics industry both technical and commercial are towards: 'open systems', compatibility, the convergence of telecommunications and computing, globalisation, trans-border takeovers and mergers, the decline of defence electronics, systems on a chip, multiprocessing, networking and the decline of the mainframe, the rise of the cordless telephone, the pre-eminence of software, transportability and technological democratisation,

Whether these trends develop or evaporate will decide the shape of the electronics industry of the Nineties.

OPEN SYSTEMS

'Open systems' or 'open systems architectures' is a phrase people bandy about in the electronics industry and it means different things to different people. What 'open systems' comes down to is trying to make the hardware and software, which together make up electronics products, interchangeable with the hardware and software of competing products.

'Open systems' basically means trying to make the hardware and software, which make up electronics products, interchangeable with the hardware and software of competing products

It that is achieved then the consumer gets the benefit of being able to choose the product which offers him the most features at the lowest price. In other words it increases competition.

Obviously it is not in the interest of established companies to have 'open systems' - they'd rather try and lock customers into their products by making it too expensive or too inconvenient for customers to buy the products of rival companies.

But if big companies are able to lock their customers into their product then they, the big companies, can dictate the pace of innovation. If, on the other hand, customers can easily go elsewhere for their products, then the pace of innovation is dictated by the fastest mover in the business.

Therefore, the 'open systems' ideal has a number of advantages: it helps to maintain open competition; it helps ensure that the best manufacturer wins in the market; and it helps to keep all the industry's products evolving and improving as rapidly as possible.

Tying the Customer to a Product

The way in which a company tries to lock its customers into its products is to try and ensure that as much as possible of the available software only works on their products. Obviously if you, the customer, can't run your favourite software program on a particular computer you are not going to buy it.

A company tries to lock its customers into its products by trying to ensure that as much as possible of the available software only works on their product

The way in which a company tries to achieve this 'lock-in' is by judicious manipulation of the three key elements which make up a computer: the microprocessor, the 'operating system' and the 'applications software'.

The way in which a manufacturer uses these three ingredients to tie customers to his products goes like this: having bought a machine the customer finds it is only useful if he has software programs (known as 'applications software') which perform functions on the machine like doing his accounts, or running his payroll; so, the customer will build up a lot of applications software to perform different tasks; then, say, he wants a new machine; so he buys one from a different company; then he finds that the new machine won't run all that applications software he has bought; so he has a choice - either he buys a whole bunch of new applications software or he buys a new machine from the original manufacturer. That manufacturer has got him.

That exclusiveness which can attach to a particular machine derives from another piece of software built into the machine called an 'operating system'. The operating system lays down the ground rules for how the applications software is written. Those ground rules can make it impossible to use on one machine the applications software produced to run on the machines of rival companies.

Furthermore the operating system can be tied to a particular microprocessor. This, as we have seen already, is the 'control chip' which controls all the other chips in an electronics product. An operating system can be designed to fit a particular design (called in the industry jargon the 'architecture') of microprocessor. So you have 'Intel architecture' micros and 'Motorola architecture' micros.

If, say, a Motorola architecture microprocessor comes along and tries to run the operating system specifically designed for an Intel architecture microprocessor either it can't, or it underperforms the Intel microprocessor by so much that it's not worth using.

An example in the real world is IBM. The operating system used by IBM's machines is called 'DOS'. DOS is designed to run on the microprocessors designed by Intel; applications software written to run on the DOS operating system won't run on anyone else's operating system.

So an IBM customer is tied in to IBM through that combination of microprocessor, operating system and applications software. The only way he can get away from IBM is to go to another manufacturer who makes machines which are 'compatible' with IBM's machines i.e. uses the DOS operating system and Intel architecture microprocessors.

Operating systems and applications software are protected by copyright - anyone using them must pay fees and royalties. Microprocessor architectures are protected by patents -anyone wanting to copy them must request and pay for a licence. Therefore the manufacturers of compatibles will have paid their dues to IBM and Intel.

So a combination of the micro, operating system and applications software can stop the customers buying elsewhere. Unless, that is, you have an operating system which can take the applications software developed for other companies' machines. Such an operating system is called 'portable'.

The Portable Operating System

A portable operating system exists in the shape of 'Unix'. Developed by AT&T in the Seventies, Unix was adopted in the mid-Eighties by AT&T, DEC, Philips, Siemens, Unisys, Ericsson, Hewlett-Packard, ICL, Nixdorf, Bull and Olivetti as a universal, portable operating system.

A portable operating system can take the applications software developed for other companies' machines

Other manufacturers have joined the Unix 'club' but with the proliferation of participants has come a proliferation of versions of Unix. This proliferation threatens Unix's potential as a truly universal, portable operating system.

Another factor in this 'open systems' trend is the advent of many new microprocessors. As we have seen, the microprocessor market has been dominated from the Seventies by the architectures of two companies - Intel and Motorola.

For the reasons given above, no other company has managed to break that stranglehold though several attempts, principally by Texas Instruments and National Semiconductor have been made. It has been said that Texas lost more money trying to establish its own microprocessor architecture than all the cumulative losses incurred in all its other product areas put together.

From the middle Seventies to the middle Eighties, Motorola and Intel shared the profits and leverage that came from their dominating position in microprocessors by licensing the designs to a number of companies for them to manufacture and sell the devices. Then in the mid-Eighties both decided to stop that policy for future micros and retain all rights to them. That seemed a big threat to the open systems concepts.

However in the Eighties the American Universities of Berkeley and Stanford developed a new concept in microprocessors known as Reduced Instruction Set Computing (RISC). This, combined with new advances in computer aided microchip design, allowed equipment makers to design their own microprocessors at a reasonable cost and so cut out the specialist microprocessor vendor.

Equipment companies like Hewlett-Packard, Sun Microsystems and MIPS Computer Systems sold systems built around such self-designed microprocessors and then licensed the designs extensively to other systems houses and also to chip manufacturers. The industry saw this as a way of breaking the Intel/Motorola domination of the microprocessor scene.

Some of the world's largest electronics companies: Philips, Siemens, NEC, Toshiba, Matsushita, Hitachi and Texas Instruments have taken licences to make these RISC micros. The Nineties will show whether this proliferation of microprocessors, with all it means for open systems, will continue or whether Intel and Motorola can re-assert their grip.

TRON - A New Japanese Operating System

Lastly TRON should be mentioned. A new operating system from Japan, TRON has been used as the basis for a new breed of microprocessors called the 'G series' from Fujitsu, Hitachi and Mitsubishi. What the effect on open systems will be of an operating system backed by the might of Japan is anyone's guess. Only the Nineties will tell.

COMPATIBILITY

Another trend of the Nineties, and a similar issue to the trend towards open systems, is a movement towards greater compatibility of equipment.

Many people first became aware of the problems of incompatibility when the first video recorders (VCRs) came out. Three different types, none of which would play each other's tapes, made the market a nightmare for the consumer. That is something the electronics industry will always try to avoid in future.

When three types of video recorders came on the market, none of which would play each other's tapes, it was a nightmare for consumers and something the electronics industry will try to avoid in the future

However there are problems. While companies and sometimes countries argue about which standard should be adopted to ensure compatibility in products - and this can take ages - there is a great temptation for a company with a product ready to be marketed to start selling it and attempt to establish a de facto standard.

If such a company is very successful in the marketplace then such a de facto standard may well be established and other companies will follow that standard in their own products and a market full of compatible products is established. If it is not successful, then other companies may come in to have a go and a fragmented market made up of non-compatible products develops.

Compatibility is a big issue in the computer business where computers are increasingly required to send data to other computers

Compatibility is a big issue in the computer business where computers are increasingly required to send data to other computers. Non-compatibility can prevent that being possible and great efforts are being made to work out standards principally by the ISO (International Standards Organisation).

In the automated factories of the future the way towards standardising the various robots and automated machinery from different manufacturers has been pioneered by General Motors. Its set of standards in the field is known as MAP (Manufacturing Automation Protocol).

In telecommunications formulating standards is clearly vital if every country in the world is to be able to communicate with every other country. But where nationalities are involved there are always horrendous problems of lobbying for one country's view or another's and this process can take so long that it seriously hinders the implementation of improvements which technical progress has made possible.

For instance the arguments over standards for the Integrated Services Digital Network (ISDN) the technique of sending voice and data down one wire at the same time, have rambled on for years. There was a ready market waiting for ISDN because it allows machines comprising combined facsimile, personal computer and telephone functions to be attached to a single telephone line. However for many years, despite having the technology to do the trick and products waiting for the starting signal, the market had to wait while the standards were thrashed out.

The answer to the standards problem is clearly to begin discussions long before the products are ready for the market. But as the pace of technological change is so quick that is easier said than done.

THE CONVERGENCE OF TELECOMMUNICATIONS AND COMPUTING

A trend that has been developing for some time which will affect the structure of the industry is the convergence of telecommunications and computing.

As the mechanical telephone exchanges are scrapped and replaced by digital exchanges which use chips instead of electro-mechanical switches, modern exchanges are indistinguishable from computers except for the software.

The structure of the electronics industry will be affected by the trend towards the convergence of telecommunications and computing

That has considerable implications for the telecommunications and computer industries. With an increasingly common technology base there will be little justification for both lots of companies to do what is essentially identical R&D and product development. That could lead to massive industry restructuring which could mean huge numbers of people from these heavily manned industries being put out of work.

GLOBALISATION

Globalisation of the industry is another trend. The big equipment companies have manufacturing plants in all the significant countries into which they sell. Since they tend, either from choice or from the insistence of the host government, to buy components locally, the big components companies set up manufacturing and marketing offices in those countries.

Increasingly the new component companies - particularly where they are involved in high margin, fast developing new markets like semiconductors - are setting up international sales offices within two or three years of the companies being started up. The modern rule for major companies and fresh entrepreneurs alike is - sell it to the world.

The modern rule for major companies and fresh entrepreneurs alike is - sell it to the world

TRANS-BORDER TAKEOVERS AND MERGERS

For companies which are not in the high margin, high growth areas of the electronics industry the only way to achieve growth is to merge with each other or take over competitors.

However there are not so many large electronics companies in Europe to allow this to happen on a domestic-only basis. So the trend is to look to other countries for mergers and takeovers and increasingly national governments are stepping aside and allowing these to happen whereas in the past they would have sought to protect their companies from a foreign predator.

The one which really set shock waves through the industry because of its audacity was when the US company ITT put all its worldwide telecommunications business into a joint venture company with the French company Alcatel (a subsidiary of the French Compagnie Generale d'Electricite) and incorporated the ITT/Alcatel business as a new company in Holland called Alcatel NV.

The merger of the Swiss Brown Boveri with the Swedish Asea and of the Swedish Ericsson with the French CGCT were similar 'trans-border' deals. The GEC/Siemens takeover of Plessey was another example.

THE DECLINE OF DEFENCE ELECTRONICS

One trend which is driving this huddling together of the large companies is the decline in defence spending. Many of the big Western electronics companies have devoted a large part of their efforts to manufacturing defence equipment for their national governments.

Large companies are huddling together partly due to the decline of defence electronics

With the new rapprochement between Russia and the West, defence spending has reduced and is expected to reduce further. Those companies which have been traditionally reliant on defence contracts are having either to look to new markets, like consumer electronics where the West is weak, or will have to merge with or takeover each other to maintain growth.

With the reduction in defence spending comes new expansive possibilities for the electronics industry. Government research establishments can feel free to release to commercial companies the results of military-funded R&D and government resources previously devoted to developing military technology can be deployed to develop industrial capability.

We are already seeing this happen in the West with programmes like Esprit and Eureka in Europe and Sematech in the USA.

SYSTEMS ON CHIPS

Another industry trend which will dominate the technology of the Nineties is the challenge to put 'systems on chips'. We have already seen how an electronics system consists of many different kinds of chip all stuck together on a PCB (printed circuit board).

These chips all perform different functions and are made with different technologies. The different functions may be memory, logic, microprocessor or linear, the different technologies may be CMOS, NMOS, PMOS, bipolar or gallium arsenide. The objective of the 'systems on a chip' brigade is to put all these functions and technologies together on one chip.

The advantages of doing this are miniaturisation, increased performance and cheapness. When, say, a memory chip has to talk to a microprocessor the signal passing from one to the other inevitably gets delayed as it travels out of one chip, along the PCB and into the other chip.

If, on the other hand, the memory and the microprocessor are on the same chip then the signal does not suffer that delay and you get significantly improved performance.

Cheapness is a considerable factor as can be seen from the prices of single chip products like calculators. When personal computers and pocket telephones can be made with single chips - and they will be in the Nineties decade - they'll become dramatically cheaper.

When personal computers and pocket telephones can be made with single chips in the Nineties, they'll become dramatically cheaper

It is expected that single-chip computers will be so cheap that shops will give away computers linked to their stores which memorise their stocks so that customers will be able to browse and order from home.

MULTIPROCESSING

Another technological change that will affect the way electronics equipment is made is the trend towards multiprocessing. This simply means using lots of microprocessors instead of, as is usual, just one or one plus a booster co-processor.

The disadvantage of the traditional approach of one micro (or one and one) is that the tasks a traditional computer is asked to perform pile up in a bottleneck going through the micro. If instead you have loads of micros 'in parallel' working simultaneously then you gain the ability to perform as many tasks at the same time as you can find micros to bolt together.

The way towards multiprocessing was set by the British government funded company Inmos which invented a microprocessor with communication links so it could interact directly with other microprocessors.

The significance of the Inmos micro, called the Transputer, is that it changed the thinking of those who build electronic systems from the traditional process of 'think of a microprocessor and build a system around it' to 'bolt a couple of micros together and add on more as and when you want the system to do more'.

The advantages can be dramatic. Using this approach you can build a computer for $100,000 which is as powerful as a $10m supercomputer, or you can put on a desktop a $10,000 personal computer/workstation, which is as powerful as a $1m mainframe computer.

> The advantages of multiprocessing can be dramatic - using this approach you can put on a desktop a $10,000 personal computer which is as powerful as a $1m mainframe computer

NETWORKING AND THE DECLINE OF THE MAINFRAME

The exploding capability of the PC leads to two more trends of the Nineties - networking and the death of the mainframe. Clearly, if you can put on a desk a machine that can do the job of a mainframe and which can communicate with other machines (the process known as 'networking'), there is little use for a big computer except as an information storage base. This is the way the industry is expected to go.

THE RISE OF THE CORDLESS TELEPHONE

Car-phones have pioneered the way with the so-called 'cellular' telephone systems organised on a country-by-country basis. This is just the beginning. Next to come along is the 'pan-European cellular' which will replace the current analogue cellular mobile phone system with a digital system which will work across Europe.

In addition, for pocket phone users, are two new cordless systems. One called 'CT2' standing for 'Second Generation Cordless Telephone' provides a one-way (makes outgoing calls but cannot receive calls), low-cost, digital telephone service which will be called 'Telepoint' in the UK. Another called 'PCN' or 'Personal Communications Network' is a more sophisticated 2-way, higher cost, digital system

THE PRE-EMINENCE OF SOFTWARE

The emergence of software as the pre-eminent enabling technology is an expected trend. By writing a software programme which instructs a machine to do a particular task, you remove the need for a human to program that machine. As software becomes more sophisticated so do the tasks which it can instruct a machine to do.

So if you have a machine as powerful and with the same memory storage capability as a brain (which we already have) you can write a software program that will instruct the machine to give the same advice as a lawyer or a doctor or a vet would give. Such programs are in their infancy because it is very difficult to write such a software program.

However, the knowledge involved in designing a car, a building, a chip or a bridge is well on the way to being automatic at the utilitarian level i.e. at the level which will work out for you how thick a steel plank you'll need to support a particular load.

This kind of machine designing ability will rise to higher and higher levels of abstractions i.e. the human designer will need to know less and less about the mechanics of bridge or car design because the computer will be making sure that all the technical considerations that will make the bridge stay up or the car run are being taken care of.

Software should be powerful enough to allow designers to draw a product on a screen all the details of building it will be immediately digitally encoded and, if so wished, transmitted to the factory for manufacturing.

That leaves the human designer free to concentrate on the aesthetics of the bridge or car. His contribution will lie not in his knowledge of how bridges or cars are built, but in his creativity. Creativity is the one thing we will never be able to automate. So in a world where everything but creativity is automated, creativity will be the key to competitive edge in the market. The most important people will be the most creative people. That is one vision of the future.

Creativity is the one thing we will never be able to automate. So in a world where everything but creativity is automated, creativity will be the key to competitive edge in the market

But before the creative types take over, the software writers will have to instruct the machines in all aspects of technical knowledge. Since this technical knowledge is rapidly expanding, it is debatable whether the software people will be able to keep on writing the programs to instruct the machines how to make the new machines which the latest advances in hardware technology have made possible.

Whatever vision of the future you choose, there can be little doubt that software will become very much more important and will account for an increasing proportion of the value added of the electronics industry.

TRANSPORTABILITY

Transportability of equipment is another major electronics industry trend. This will be seen particularly in goods which need a screen like TVs and personal computers (PCs). The bulkiness of these derives from their use of glass cones enclosing a vacuum to form the display or screen (called cathode ray tubes or CRTs); during the Nineties these will be replaced by the same type of display as you get in a digital watch or a calculator i.e. a flat, light affair made from liquid crystal (an lcd or liquid crystal display).

Other equipment which is shrinking rapidly is the video camera - the camcorder - which appears to halve in size and price about every five years. When these are fitted with communication links allowing them to send images by mobile telephone techniques then every citizen will be their own one-man television camera crew - reporting what is happening, wherever he or she is, whenever he or she likes, to whoever he or she likes.

A key trend of the Nineties will be that all kinds of equipment will become transportable and usable anywhere in the world as it is linked into the world's communications systems

Mobile telephones are shrinking in size and weight so rapidly that they will become the size and price and as commonplace as calculators during the Nineties. Another shrinking piece of kit is the facsimile machine which is reducing in size and price at much the same rate as the video camera and is becoming more transportable using mobile telephone links.

So one of the key trends of the Nineties will be that all kinds of equipment will become transportable and usable anywhere in the world as it is linked into the world's communications systems.

DEMOCRATISING EFFECT OF TECHNOLOGY

The transportability of electronics equipment and its ability to communicate worldwide will fuel another trend in the electronics industry - a trend towards the democratisation of technology.

Dictatorships depend on the suppression of truth. They can't survive in an environment of free exchange of information. When TV cameras can be carried in a pocket and their pictures can be beamed by satellite to anywhere in the world it will become impossible to suppress truth. The TV cameras in Vietnam and Tiananmen Square told the world the truth and governments were unable to dismiss the atrocities as inventions and lies.

When Joe Citizen carries his own medium for global exchange of information in his pocket, global media are not going to be manipulable by governments. It will be impossible to falsely persuade the citizens of one country that they can't get a better deal in another country.

The fact that Joe Citizen is his own global communication medium will mean that the 'official' media - newspapers, TV, radio which are controlled and manipulated by tycoons and governments - will have to tell the truth or become hopelessly discredited. The shrinking transistor will shrink the world into a village where word of mouth will pass the truth to everyone.

The shrinking transistor will shrink the world into a village where word of mouth will pass the truth to everyone - the 'official' media will have to tell the truth or become hopelessly discredited

CHAPTER 10

WHAT'S IN IT FOR ME?

How to become a millionaire; the developments
in the key component technologies for the
Nineties - chips, screens, batteries and CD-
ROMs. What the technology/price trends are;
how to predict them; how to exploit the 'learning
curve'.

HOW TO BECOME A MILLIONAIRE

Do you want to be a millionaire? Well the new entrepreneur millionaires of the Nineties will be the ones who can correctly predict the trends in components technology and can match a new product idea to the latest capability in components.

The new millionaires will be those who can predict the trends in components technology and can match a new product idea to the latest capability in components

The Technology Trends of the Key Components

The general technology trends of the key components are obvious enough. For instance:

- we know that chips are going to hold more information for less cost, that they are going to be able to process more information for less cost and that they are going to work faster;

- we also know that flat LCD screens are going to get bigger, better, and cheaper;

- we know that batteries are going to get lighter, more powerful, longer-lasting and increasingly re-chargeable;

- we know we'll have new memory storage techniques like CD-ROMs (Compact Discs which store information rather than music) which can store the contents of the Guinness Book of Records including animated pictures on a 4-inch diameter disc.

WHAT THE TECHNOLOGY/PRICE TRENDS ARE AND HOW TO PREDICT THEM

The clever part is predicting when the major advances in each component area are going to occur and when mass production of a new advance is going to make the component available sufficiently cheaply for you to put it into a product which people can afford to buy.

Sometime in the Nineties it may be possible to put together the latest examples of all the components mentioned above and get something the size and weight of a paperback book with a colour screen on one side and a keyboard on the other which can search through and display the equivalent of the contents of an encyclopaedia.

If you were to make this particular tool in 1990 it would cost upwards of £5,000. If you make it in the year 2000 it will probably cost a few hundred pounds and that's the price at which you tend to get the mass market sale.

If you wanted to be the manufacturer of this little tool you would have to persuade the components companies, which produce CD-ROMs, microprocessors, memory chips, flat screens and the latest batteries, to let you see their future plans. The components companies would also need to give you firm commitments on the delivery of each ingredient, and to time the introduction of your tool for when the components companies are mass-producing the latest types of each of these components and can consequently supply them cheaply.

That is easier said then done. In the home computer boom of the early Eighties, many entrepreneurs went broke because, although they designed good products which everyone wanted, they were unable to keep supplying computers to the market because the supply of components ran out.

A further snag is that contracts in the chip business are traditionally subject to what is politely termed 're-negotiation'. If prices slump then equipment manufacturers return chips or cancel orders for chips which they have contracted to buy. If, on the other hand, shortages produce high prices, chip manufacturers often force equipment makers to pay higher prices for contracted supplies.

Another problem for equipment manufacturers is that the expectations of the components companies on when they would have supplies of new technology components has been over-optimistic, sometimes wildly over-optimistic, so the pitfalls involved in trying to make latest technology products are considerable.

Equipment manufacturers face many problems including over-optimistic claims from components companies over when they can supply new technology components

Therefore you might decide it's simpler to make cheaper versions of old products for which there's a well-tested and plentiful supply of the necessary components. To find out what you should be copying, you only need to look at what kinds of products are being bought by companies for their professional staff for £3,000 to £50,000. It would, on past trends, be sensible to expect those sort of products to be in the shops for a few hundred pounds within five to ten years.

At the beginning of the Nineties, for instance, design engineers are just beginning to use personal computers/ workstations (the terms are becoming interchangeable) that will allow them to do designs using 3-D images on-screen. (Until now, the computing power needed to do that is so great, people have had to use super-computers).

At the beginning of the Nineties, a 3-D machine like this will cost £50,000. By the end of the Nineties, you'll probably be able to buy it for a few hundred pounds. The first entrepreneur to have it on the market for that kind of price will make tens of millions and the second and third guys will probably make millions.

These entrepreneurs will have monitored the dropping component prices, arranged for sufficient supplies when they judge the time right, and will have set up assembly factories to leap into action and put all these components together to make 3-D computers at the first possible moment. So the skill of the would-be millionaire in this situation is in organising and predicting: organising consistent supplies, predicting when the price of all the key components to make up a product is low enough to make the cost of producing that product attractive to the consumer.

> **The skill of the would-be millionaire is in organising consistent supplies and predicting when the price of the key components which make up the product is low**

DEVELOPMENTS IN KEY COMPONENT TECHNOLOGIES

THE CHIP AND ITS LIFECYCLE

So let's look in more detail at the components. The key components in computers are still the chips. The chip that holds most memory and is therefore the most widely used is the DRAM (Dynamic Random Access Memory). The rule of thumb with DRAM is that a new 'generation' comes out every three years but stays in production for about ten years. So there's a lot of overlap between the rising generation, the peak generation and the declining generation. Each new generation quadruples the storage capacity of the previous generation.

The DRAM generation which peaks in 1990 is the million-bit DRAM known as the 'Mega-bit' or 'M-bit'. It has sufficient storage capability to hold 40 pages of A4 size typing paper with 54 lines of single spaced typescript on each.

Predicting the Chip 'Learning Curve'

When the Megabit first came on the market in 1987 it would cost you $200 or more. In 1988 you could have paid anything from $25 to $50 depending on whether you bought under contract from a manufacturer or from a dealer at the height of the scarcity.

By contrast, in 1989 you could buy Megabit DRAMs for $10. Prices traditionally bottom out during the peak year for each generation of the device. For the Megabit that's 1990 when there's the maximum number of suppliers each producing at maximum capacity although late 1989 saw suppliers reducing their output in an attempt to keep prices up.

The rule of thumb is that the price of all chips goes down to about the $5 level, though in one of the worst recessions it was possible to buy four DRAMs chips for $1. Right at the end of their life cycle their price tends to rise as suppliers drop out and shortages occur.

The rule of thumb is that the price of all chips eventually goes down to about the $5 level

1989 saw the first examples of the next generation of DRAM hitting the market. This is the 4Mega-bit chip, holding 160 pages of information and, costing about $200 at the beginning of 1989 and expected to be $40 in mid-1990.

Under the industry's three-year cycle rule, the 4 Mega should have its peak market in 1993 when it, in its turn, will be selling for around the $10 mark having seen a price decline through 1990/91/92.

So an entrepreneur who planned for a luxury, advanced technology product offering the maximum memory (PCs usually have eight DRAMs each) would have planned to launch an expensive top-of-the-range model in 1988 using Megabits. And he would be planning to bring out his next upgraded model using 4Mega chips in 1991.

But an entrepreneur wanting a cheapie me-too kind of product using Megabits would probably wait till the end of 1990 to launch his product. That way the cost of his DRAM would be about one third or less than the cost of 1988 DRAM.

The other chips used in the 1988 model could be integrated together into a much smaller overall number for the 1990 model (half the number every year), and the resulting cost of the PC would be maybe a quarter or less than the cost of the 1988 model. Chips represent 30 to 40% of the final price of this kind of product.

So that's the chip 'learning curve', as this price decline process is called, in operation. After the 4Mega we should see the introduction of the 16Megabit (640 A4 pages) in 1992 (peak market 1995), the 64Mega in 1995, the 256Mega in 1998 and maybe the first sightings of the 1Giga to celebrate the year 2000.

A New Development in Memory Chips - Flash E2PROMs

A new development in memory chips may make possible totally solid state personal computers - i.e. computers with no moving parts. In practice that means computers without discs.

At the moment the memory chips you use in a computer are DRAMs which lose their memory when you turn off the power. So, if you want to keep the information you've put in the chips, you transfer it to a disc before turning off the power.

However in the Nineties, chips which are denser yet just as cheap as DRAMs and which do not lose their memory when the power is turned off are on the way. These chips are called 'Flash E2PROMs'.

The beauty of using Flash chips rather than discs for a portable computer manufacturer is that chips don't have to have a revolving mechanism so there's no need for a motor and therefore no need for hefty batteries to power a motor. That means less weight and a smaller size.

At present Flash chips are in commercial production at the Megabit stage though 4Mega versions are around in laboratories. The leading producers of Flash, Intel and Toshiba, are saying that 16 Mega versions of Flash chips may be around before 16 Mega versions of DRAM.

As we have projected above, 16 Mega DRAMs should first appear in 1992 with a peak market in 1995. That means we may have 16 Mega Flash chips around in 1991. With 2Megabytes of storage (1 byte = 8 bits) they would be more than powerful enough to take over from discs which typically store either 720k-bytes (single density) or 1.4Megabytes (double density).

The only snag is that floppies cost £1 to £2 whereas chips start at a very expensive price and don't drop to £2 until they've been on the market for four or five years.

So some smart entrepreneurs will be scratching their heads working out the advantages and disadvantages of chips versus discs and figuring out when the price will be right to make the swap. The big problem for the entrepreneur is the problem of 'When?'. Timing is everything!

CHANGES IN MICROPROCESSORS

Another change in the chip world is going to be in microprocessors. A new breed of microprocessors developed at the American universities of Berkeley and Stanford called RISC (Reduced Instruction Set Computing) processors have been freely licensed around the industry and are on the way in large quantities. They should have two main effects:

A new breed of microprocessors called RISC processors have been freely licensed around the industry and are on the way in large quantities

- they are very fast and should make it possible for machines to do more things than they now do;

- as there are unprecedentedly large numbers of companies able to make them there will be fierce competition and they will therefore be plentiful and cheap.

The result is likely to be that the entrepreneur is going to be able to get his hands on very powerful microprocessors very soon after their introduction and very much cheaper than ever before. This means that the small guy gets the same shot at the market as the large guy.

THE REPLACEMENT SCREEN

After the chips the next most important component in electronic equipment is the screen or 'display' as it's called in the electronics industry. Here the main aim is to develop a replacement for the bulky, heavy, high power-consuming glass pyramid called a cathode ray tube which is used in nearly all TVs, computers and instruments.

The expected replacement is the liquid crystal display or 'lcd' which is flat, low power consuming, light and relatively cheap to make. This is the type of display used in watches and calculators. At such a small size they can be churned out very quickly.

Electronics companies have a Holy Grail in their lcd development - to produce a colour lcd 206mm by 156mm which can handle 60 lines of text

The electronics companies now have a Holy Grail in their lcd development - to produce a colour lcd size 206mm by 156mm which can handle 60 lines of text (like a cathode ray tube).

That is the component which will allow the making of a Nineties dream product - the flat screen, paperback book sized, light-weight portable computer. How far away is such a display? No one knows. Many companies can show you one, but no one is claiming to have a factory which can turn them out in large enough volumes to bring the price down to a viable level.

In all components manufacturing the key question is not just 'Can you make it?' but 'How much can you make it for?' As elsewhere, manufacturing volumes drive unit costs. And overhead costs are vital where the component is to be used in applications which are semi-consumer - on the borderline of business and personal use.

The Likely Price for the LCD

One clue to the likely price can be seen in the black and white, or 'monochrome' as they call it in the industry, displays; £70 is currently being charged for a screen capable of displaying 30 lines of text (231mm by 105mm). For 60 lines the price is around £100. That's about £40 for the screen and £60 for the electronics circuitry needed to make it work. A colour version would probably be made in volume for an overhead cost not dissimilar to this.

But will it be and when will it be? Those, for the entrepreneur, are the key questions. For the moment, the manufacturers are making good profits out of monochrome displays and expect prices to remain stable.

For how long? The only answer is to say 'Look what happened in the watch display business'. There the prices were high when only the Japanese could make the displays. As soon as other countries could make them, prices dropped like a stone.

At the moment, only the Japanese can make these big displays of either 30 or 60 lines. They have learnt that to extract the most profit from a product you delay the introduction of a better product for as long as you can. But you can only do that when no one else can make the better product. So, while the Japanese continue to own this bit of the components industry it is probably unrealistic to expect prices to drop.

The shrewd entrepreneur of the Nineties will therefore be monitoring the capabilities of the West and will be timing the launch of his products using large flat screen displays for a year or two after the West is able to manufacture them. That's exactly the moment when the Japanese will flood the market with incredibly cheap flat screen displays.

IMPROVEMENT IN THE BATTERY

Another vital ingredient of the new increasingly mobile and pocketable products of the Nineties is the battery. Here the technological improvement is in one direction - batteries that will give out more power for longer in a smaller, lighter pack.

That will most likely be a process of gradually evolving improvement which is a dangerous statement to make in the electronics industry. For instance a late 1989 apparent breakthrough at the Bell Labs-derivative 'Bellcore' announced batteries had been developed that were 50 microns thick using silicon and molecules of water. A human hair is 30 microns thick.

If the Bellcore battery, or someone else's equivalent, becomes a commercial reality then all bets are off and the portable equipment industry will be on a new roller-coaster.

A lot of battery companies will be focusing on getting batteries to the point where they are genuinely small enough, light enough and powerful enough for pocket telephones which is seen as a major market.

New materials like polymers will also be developed for use in batteries. These are likely to allow batteries to churn out a higher charge over a longer period. These improvements are expected to be gradual and undramatic.

Rechargeability

However in one area there is expected to be a lot of money spent - and that usually leads to rapid improvement in capability - that area is rechargeability. The reason why a lot of resource will be devoted to it is because rechargeability of batteries is very much on the Green agenda.

That's because batteries use heavy metals like cadmium, mercury and lead which are massive pollutants when disposed of. Since their use cannot be much reduced in current batteries, the only way to stop less of these metals being dumped is to reduce the number of disposable batteries. The only way to do that is to boost the use of rechargeables.

There is expected to be a lot of money spent on rechargeability - as the rechargeability of batteries is very much on the Green agenda

The only thing stopping the more common use of rechargeables is their power relative to disposables, their price and the number of recharging cycles they permit.

The power differential is something the technology will gradually improve on. The recharging cycles can be up to about 1000 times and will be increased - new materials could mean a big improvement. The price differential is merely a function of manufacturing volume. As rechargeables become more and more in demand and the manufacturers increase production to meet it, prices will drop to the level of disposables.

The battery issues - unless there's a major breakthrough of the Bellcore type - may not have a dramatic effect on the kind of electronics goods that can be made, but CD-ROMs could.

CD-ROMs AND THEIR FUTURE

CD-ROMs increase the amount of storage capability on a 4-inch disk by a vast amount. One disc can hold 550 M-bytes - enough for the Oxford English Dictionary. The discs can also store pictures, animated pictures and speech.

That could mean, for instance, that you can have a screen in your car on which you can display a map of any area accompanied by a voice-over telling you when you're coming up to a turn with a picture of the turn so you recognise it.

Or you can have a few books on a disc, or a teach yourself course in karate with animated pictures demonstrating falls and grips, or a language course combining text with spoken examples. CD-ROMs could also become the storage medium for portable automatic translation equipment or automatic speech recognition equipment both of which require vast amounts of storage.

So when will these things be around in the shops? There are signs that the producers are wary of just selling the discs and equipment on which people can put whatever they want on the discs. Having been bitten by the tape market where people bought blank tapes and recorded what they want on them, manufacturers are looking for ways to maximise profits.

They are doing that by selling already programmed CD-ROMs for vast amounts of money. For instance if you want the Oxford English Dictionary on a CD-ROM it will cost you £500; if you want the Grolier Electronic Encyclopaedia on CD it will cost you £230. Yet a CD-ROM costs the same to make as a Compact Disc (CD) and those are being sold for £5 or less.

In 1988 the technology to make erasable reprogrammable CD-ROMs was invented - but as yet no one is willing to manufacture them at volumes which will get the price down

There are signs of price breaks - you can buy inter-active games using CD-ROMs for £30 or so, but what the consumer is waiting for is the CD-ROM which he can erase and re-program himself like a floppy disc, a tape or a chip.

The technology to make erasable reprogrammable CD-ROMs was invented in 1988. There is no problem with it, but there is no one as yet willing or able to manufacture them at volumes which will get the price down.

As with batteries, chips and displays, the cost is volume-driven. Once some clever entrepreneur finds a use for them which guarantees him huge sales, he will doubtless be able to persuade a manufacturer to churn out reprogrammable CD-ROMs at CD prices.

As always, the advances in component technology continue to drive the equipment industry and, as always, the people who control the hardware manufacturing still control the pace at which new products using new technology are introduced to the market.

Where the technologies get into the hands of a few people who are bent on maximising profits at any cost, then the power exists to stifle all entrepreneurial activity. Only by widely diffusing the technology are new companies able to move into action, and only by new companies coming in do you get true competitiveness.

A future challenge is to keep electronics technology diffused, the entrepreneur active and competition alive

One of the challenges of the Nineties will be to keep the technology diffused, the entrepreneur active and competition alive.

CHAPTER 11

WHERE WILL IT ALL END?

Will the shrinking transistor continue as the main force? Will superconductivity take over? Will bio-technology take over? Will Japan take over the world computer industry? What would be the implications for entrepreneurism? Big business versus individuality. Can the US regain technological superiority? Can Europe re-build its technological and industrial strength? Will Eastern Europe become a significant market/producer?

The incredible shrinking transistor has been the driving force of the electronics industry for the past 40 odd years. Will it continue to be?

Alternatives to silicon technology pop up from time to time:

- in 1987 it was superconductivity;

- bio-technology - genetically engineering bugs to produce an equivalent to a chip - is another alternative which is talked about;

- optical computers - which use strands of glass down which pulses of light run as an alternative to electrons flowing through metals - are also talked of as an alternative to silicon;

- for years some people have been predicting that gallium arsenide is the technology of the future (silicon technologists add 'and it always will be').

But although it has often been predicted that silicon technology will hit a wall and that further shrinks will be impossible, the technology has kept on evolving from the 8 micron line widths of the early Seventies to the 1 micron line widths used in mass production chip factories in 1989.

Already we're seeing small scale commercial production at 0.8 micron at which level it is possible to make 4 Megabit DRAMs. Laboratory specimens of 16 Megabit DRAMs using 0.5 micron are in existence. The leading commercial research laboratories are working on 0.35 micron manufacturing processes on which the 64 Megabit DRAM will be made.

There seems no end to the shrinking transistor for a long time to come unless there's some dramatic breakthrough in one of the other technologies

In university and research laboratories X-rays have successfully made chip line widths less than a tenth of a micron. There seems no end to the shrinking transistor for a long time to come unless there's some dramatic breakthrough in one of the other technologies.

WILL SUPERCONDUCTIVITY TAKEOVER?

Superconductivity could be the one. What it means is the phenomenon of electrons flowing through materials without meeting any resistance. When that happens everything they do is speeded up dramatically.

Superconductivity is the phenomenon of electrons flowing through materials without meeting any resistance

The British Nobel-prizewinner Brian Josephson first noted that when certain materials were cooled to absolute zero the phenomenon of superconductivity occurred. Chip companies still make experimental devices (known as 'Josephson Junctions') based on his theories, but the necessity for freezing them to near absolute zero (-273 degrees C) - only achievable by immersing them in liquid helium, which is difficult and expensive - has always ruled them out as a commercial technology.

Then came 'The Woodstock of Physics' as the March 1987 meeting of the American Physical Society in New York was dubbed. The meeting erupted in cheering and applause as the Universities of Houston and Alabama disclosed that they had made materials (compounds of yttrium, barium, copper and oxygen) which allowed superconductivity at around -180 degrees C.

This meant that instead of having to use liquid helium to cool the circuits sufficiently for them to superconduct it was possible to use liquid nitrogen which, as one scientist said, is 'as cheap as beer and as plentiful as air'.

For a while after 'Woodstock' the scientific world went crazy as new claims were made from all over the globe. The Americans, in their traditional way, formed venture capital backed start-up companies to exploit the new technology. One, called Conductus, staffed by academics from California's Stanford and Berkeley Universities, produced chip devices called 'Squids'.

But when the claims from scientists of further moves up the temperature scale came to a halt, the immediate industrial hopes for superconductivity dried up. Although liquid nitrogen might be a more practicable coolant than liquid helium, the cooling problem still ruled out any widespread commercial exploitation of the new materials.

If room temperature superconductivity could be achieved, it would revolutionise the electronics industry allowing chips to be made that would work between 1,000 and 10,000 times faster than today's chips

However the possibility of a new superconductivity breakthrough always exists. Room temperature superconductivity would revolutionise the electronics industry allowing chips to be made that would work between 1,000 and 10,000 times faster than today's chips.

THE BUG AS AN ALTERNATIVE

Another alternative to the silicon chip could be the bug. It may be possible to genetically engineer bugs which could then produce devices which would replace chips. Clearly if you can grow chips in a vat (like yeast and sugar working to produce alcohol) - instead of in a $400m silicon chip factory - then you would get dirt-cheap chips.

Already ICI has announced that it has genetically engineered bugs which can make plastic. The next step would be to get a bug which can make electro-active polymers (plastics which conduct electricity).

And after that you'd engineer a bug which could produce electro-active polymers in the shape of electronic circuits. At that stage you would be able to manufacture chips by stirring up some bugs with a few materials in a bucket. The prospects are mind-boggling. Some scientists reckon such circuits will be made in the Nineties.

OPTICAL COMPUTING

Optical computing has advantages in certain uses - for instance where it is necessary to manipulate light signals as in artificial vision in robots - analogue optical computing cuts out the complicated programming which a silicon digital computer would need to perform the same tasks. But optical computing has not got enough advantages over silicon-based computing to justify the massive investments required to get it to the point where it could challenge conventional computers for the mass market. It is expected to remain as a niche activity for specialised uses.

Gallium arsenide has many advantages over silicon being theoretically six times faster, radiation resistant (which silicon isn't) and less power consuming. However, it is difficult to process and the sheer extra cost of manufacturing with gallium arsenide rather than with silicon is likely to rule it out as a mainstream technology.

Silicon continually surprises technologists with its ability to 'stretch' (as they call it) and its capabilities to produce more speed and higher performance. For the foreseeable future it looks irreplaceable as the key enabling technology for the electronics industry. Any country which can dominate the world supply of chips is going to have its hand on the throat of the world electronics equipment industry of the Nineties.

For the foreseeable future silicon looks irreplaceable as the key enabling technology for the electronics industry

WILL THE DOMINATION BY JAPAN CONTINUE?

Japan is already producing over half the world's chip output.

The outcome of the Chip Wars of the Nineties will give the clue to who is winning in the electronics industry and in the wider international economic struggle. The chip war is just a preliminary round to the main event.

In the Chip Wars, the West is currently suffering from schizophrenia. There is already a battle going on between those in the USA and Europe who want protection for the local chip industries via subsidies and tariffs and those who want all tariffs removed so they can buy their chips as cheaply as possible.

The West is in a very difficult position. If it applies tariffs to protect its local chip industry it may kill off the local equipment industry which will be unable to get chips as cheaply as the Japanese.

But if the governments of the West allow the Japanese to sell chips to the West's equipment makers at any price then the West's chip makers will be put out of business. And once they're out they make not be able to get back in.

That's because the chip industry depends on those who make the chip-making equipment. If you kill off your national chip industry, you kill off your national industry for chip-making equipment. When that's happened you'll never get back into chip-making without unbelievably huge investment.

The demise of the West's chip industry, it is argued, will lead to the West's equipment industry being strangled later on when the Japanese, at their leisure, can deny the West chip supplies knowing it has no alternative means of supply.

The demise of the West's chip industry could lead to the West's equipment industry being strangled if the Japanese decide to deny the West chip supplies

There is one current alternative source of supply - Korea - but it is only supplying something like 2% of the world's requirement for commodity chips (though it's growing at a rate of between 150% and 200% a year). Another expected source of supply - mainland China - is still some way away (it is thought) from being able to supply the latest commodity chips. Taiwan has stayed away from the commodity chip business.

So, if the West loses its chip industry and has no alternative means of supply, the next thing to go will be the West's computer industry. If the Japanese take that, it is thought, they may not adopt the same attitude to open systems and compatibility as the West has done.

As we've already seen, by judicious use of the microprocessor and operating system it is possible to lock your customers into your product. It is possible that the Japanese would want to do that. If they did, the computer industry would be dominated by Japanese Big Business.

Without adherence to principles like open systems, standardisation and compatibility, competitive entrepreneurial activity would be stifled. If it is, then the large Japanese companies would dictate the pace at which technological advance is adopted and the price everyone else pays for it.

From there it is but a short step to the West losing most of its remaining electronics industry. Without an electronics industry the West will be reduced, as Bob Noyce of Intel puts it, 'to taking in Japanese laundry'.

NEW POTENTIAL ELECTRONICS MARKETS FOR THE WEST

Which would be a pity because the West is opening up vast new potential electronics markets. For instance:

- governments are deregulating telecommunications equipment markets which always provides for more competition and expanded markets;

- governments are licensing a lot of people to run mobile communications networks which will stimulate highly competitive markets for portable telephones;

- many countries are installing fibre-optic networks which will open up a big demand for data transmission equipment;

- Eastern Europe is opening up to the West and showing itself to be both hugely hungry for consumer goods and stuffed with personal savings to spend on them. (It remains for governments to work out how they spend those savings which are in non-convertible currencies).

It would be very silly if the West was to create all these market opportunities and then fail to profit from them because it was unable to produce the goods to sell into them. Which is why a lot could depend on the decisions which the West takes in the early Nineties about its chip industry.

Options for the West

Basically the West has three options:

- to go for a totally regulated trade in chips under which every trading block has to import as much as it exports;

- to go gung-ho for free trade;

- to carry on with the policy of free trade softened by subsidies, tariff protection and preferred local procurement.

The option it chooses could decide the future of the West's electronics industry. The newspapers of the early Nineties should be interesting as they unfold a tale which could have a profound effect on the economies of the West.

THE GOLDEN AGE OF ELECTRONICS?

At the moment the conditions exist for a golden age of electronics as it sits on the verge of becoming the world's largest industry. It could provide the human race with tremendous opportunities for leisure and pleasure, for cheap readily available global communication and information exchange, for unprecedented entrepreneurial activity, for automated mass-production of cheaper and cheaper goods, for a flowering of creative activity and for increased worldwide democratisation.

At the moment the conditions exist for a golden age of electronics as it sits on the verge of becoming the world's largest industry

But conditions can change. It will, very largely, be up to the regulators and the authorities worldwide to ensure that these opportunities do not lead to any one country or group of companies dominating the electronics industry to the extent that they can dictate the pace of technological change, the directions in which technology moves, the conditions under which the technology industry operates and the prices which are charged for technology products.

The free market, entrepreneurial, cut-throat operating style of the electronics industry has forced the companies to keep innovating, developing technology and charging low prices - in the Nineties, the challenge will be to keep things that way

The free market, entrepreneurial, cut-throat operating style of the electronics industry has forced the companies to keep innovating and developing the technology and charging rock-bottom prices for it. That has benefited us all. In the Nineties, the challenge will be to keep things that way.

INDEX

COMPUTER WEEKLY PUBLICATIONS

Computer Weekly is the UK's leading weekly computer newspaper which goes to over 112,000 computer professionals each week. Founded in 1967, the paper covers news, reviews and features for the computer industry. In addition, *Computer Weekly* also publishes books relevant to and of interest to its readership.

Publications to date (obtainable through your bookshop or by ringing 081-685-9435/081-661-3050) are:

The Computer Weekly Guide to Resources 1990

Our extensively indexed second Annual Guide fulfils the computer industry's need for an independent, handy, up-to-date reference review signposting and interpreting the key trends in the computer industry.

A key section this year is an in-depth independent discussion of 270 software and computer companies, invaluable for managing directors, DP managers, sales and marketing people and all executive job hunters.

Our first Annual Guide was well acclaimed: 'In spite of a plethora of guides to various aspects of the computer industry, there hasn't been one readable, comprehensive overview of the current UK scene. *Computer Weekly*'s Guide to Resources has filled the bill ... it's very good.' *The Guardian*

ISBN 1-85384-017-3   416 pages   A4 size   Price £45.00

Aliens' Guide to the Computer Industry by John Kavanagh

In a lucid and light style, leading computer industry writer John Kavanagh discusses how the various parts of the computer industry inter-relate and what makes it tick. Complete with extensive index, the book is invaluable for all who come into contact with the computer industry.

'Business professionals who worry about their grasp of the general computing scene and do not want to be bombarded with jargon and technicalities, will get good value ... an excellent 'snapshot' of the companies, the current areas of interest and the problems' *Financial Times*

ISBN 1-85384-012-2   192 pages   A5 size   Price £9.95

Computer Jargon Explained by Nicholas Enticknap

Following reader demand this is a totally revised, expanded and updated version of our highly successful guide to computer jargon, *Breaking the Jargon*.

This 176 page book provides the context to and discusses 68 of the most commonly used computer jargon terms. Extensively cross-indexed this book is essential reading for all computer professionals, and will be useful to many business people too.

'... a useful shield against the constant barrage of impossible language the computer business throws out' *The Independent*

'... a worthwhile investment' *Motor Transport*

ISBN 0-85384-015-7   176 pages   A5 size   Price £9.95

What To Do When a Micro Lands On Your Desk
by Glyn Moody and Manek Dubash

This book will help you get the most out of your microcomputer. It is a practical book, giving advice on how to make the transition from typewriter to micro profitably and with minimum effort.

The authors look at software - wordprocessing, databases, spreadsheets, graphics and communications - and the different types of hardware on the market. The book contains valuable information on training, health and security, and legal matters including the Data Protection Act, operating systems, the history of the computer, the current micro scene and the future.

ISBN 1 85384 011 4   296 pages   A5 size   Price £14.95

Open Systems: The Basic Guide to OSI and its Implementation
by Peter Judge

We recognise the need for a concise, clear guide to the complex area of computer standards, untrammelled by jargon and with appropriate and comprehensible analogies to simplify this difficult topic. This book, a unique collaboration between *Computer Weekly* and the magazine *Systems International*, steers an independent and neutral path through this contentious area and is essential for users and suppliers and is required reading for all who come into contact with the computer industry.

ISBN 1-85384-009-2   192 pages   A5 size   Price £12.95

IT Perspectives Conference: The Future of the IT Industry

ISBN 1-85384-008-4   224 pages   A4 size   Price £45.00

Profiting From Your Printer: Users' Guide to Computer Printing
by Frank Booty

ISBN 1 85384 019 X   about 160 pages   A5 size   Price £14.95

A Simple Introduction to Data and Activity Analysis
by Rosemary Rock-Evans

ISBN 1-85384-001-7   272 pages   A4 size   Price £24.95

Computer Weekly Book of Puzzlers Compiled by Jim Howson

ISBN 1-85384-002-5   160 pages   A5 size   Price £6.95

Considering Computer Contracting? by Michael Powell

ISBN 1-85384-022-X   176 pages   A5 size   Price £12.95

Women in Computing by Judith Morris

ISBN 1-85384-004-1   128 pages   A5 size   Price £9.95

How to Get Jobs in Microcomputing by John F Charles

ISBN 1-85384-010-6   160 pages   A5 size   Price £6.95

Low Cost PC Networking by Mike James

ISBN 0-434-90897-5   256 pages   246 x 188 mm   Price £16.95

Selling Information Technology: A Practical Career Guide
by Eric Johnson

ISBN 0-85012-684-3   244 pages   144 x 207 mm   Price £12.50